Information Systems Engineering and Management

Volume 44

The book series "Information Systems Engineering and Management" (ISEM) publishes innovative and original works in the various areas of planning, development, implementation, and management of information systems and technologies by enterprises, citizens, and society for the improvement of the socio-economic environment.

The series is multidisciplinary, focusing on technological, organizational, and social domains of information systems engineering and management. Manuscripts published in this book series focus on relevant problems and research in the planning, analysis, design, implementation, exploration, and management of all types of information systems and technologies. The series contains monographs, lecture notes, edited volumes, pedagogical and technical books as well as proceedings volumes.

Some topics/keywords to be considered in the ISEM book series are, but not limited to: Information Systems Planning; Information Systems Development; Exploration of Information Systems; Management of Information Systems; Blockchain Technology; Cloud Computing; Artificial Intelligence (AI) and Machine Learning; Big Data Analytics; Multimedia Systems; Computer Networks, Mobility and Pervasive Systems; IT Security, Ethics and Privacy; Cybersecurity; Digital Platforms and Services; Requirements Engineering; Software Engineering; Process and Knowledge Engineering; Security and Privacy Engineering, Autonomous Robotics; Human-Computer Interaction; Marketing and Information; Tourism and Information; Finance and Value; Decisions and Risk; Innovation and Projects; Strategy and People.

Indexed by Google Scholar. All books published in the series are submitted for consideration in the Web of Science.

For book or proceedings proposals please contact Alvaro Rocha (amrrocha@gmail.com).

Sanjay Taneja · Swati Gupta · Mohit Kukreti ·
Ercan Ozen
Editors

Navigating the Deepfake Conundrum: A Manager's Roadmap

Springer

Editors
Sanjay Taneja
Department of Management Studies
Graphic Era (Deemed to be University)
Dehradun, Uttarakhand, India

Swati Gupta
Chitkara Business School
Chitkara University
Punjab, India

Mohit Kukreti
Department of Business Studies
University of Technology and Applied Sciences
Ibri, Oman

Ercan Ozen
Department of Finance and Banking
Uşak University
Uşak, Türkiye

ISSN 3004-958X ISSN 3004-9598 (electronic)
Information Systems Engineering and Management
ISBN 978-3-031-90741-8 ISBN 978-3-031-90742-5 (eBook)
https://doi.org/10.1007/978-3-031-90742-5

This Springer imprint is published by the registered company Springer Nature Switzerland AG
The registered company address is: Gewerbestrasse 11, 6330 Cham, Switzerland

Foreword

The fast technological progress in artificial intelligence delivers cross-sector innovations together with unique and unanticipated challenges. The intersection of benefits and dangers from deepfake technology positions it between advancement possibilities and severe threats toward security along with ethical guidelines and trust integrity.

Managers and decision makers face an urgent requirement to grasp and handle disruptions resulting from deepfakes because businesses need to adapt to this transforming digital environment. The technology behind deepfakes transforms fundamental aspects of business management including company reputations and confidence levels of investors and established practices of transparency and customer loyalty.

Professionals need an essential guide in the form of "Navigating the Deepfake Conundrum: A Manager's Roadmap" to understand deepfake technology in various sectors. This study makes its argument through data science methods such as Bayesian analysis paired with structural equation modeling along with regression techniques for delivering empirical findings about deepfake risks and opportunities.

Through the connection between academic discoveries and practical business uses this publication delivers operational guidance to organizations regarding their ability to increase their resilience levels while improving transparency and establishing ethical protection measures. The roadmap acts as a critical resource that benefits corporate leaders as well as policymakers and researchers in their journey through the deepfake-controlled digital age.

I commend the authors for their **visionary work** in shedding light on this critical issue and providing **managers with the tools they need to stay ahead** in an AI-powered future.

Prof. (Dr.) Sanjeet Kumar
Head and Professor
Department of Business Administration
Chaudhary Devi Lal University
Uşak, Türkiye

Preface

Deepfake technology enhancements have created widespread industrial disputes which involve business operations alongside financial institutions along with media organizations and government structures. Its innovative capabilities undeniably transform marketing strategies as well as entertainment and education but deepfakes create major obstacles for maintaining corporate transparency while preserving ethical decisions and maintaining stakeholder trust. Modern-day digital reality faces critical challenges since AI achieves precision-based modifications to both audio and visual files and imagery.

Business leaders together with managers alongside policymakers need to develop deepfake risk identification skills and protection capabilities which help understand this technology and reduce its dangers. The authors wrote this book *Navigating the Deepfake Conundrum: A Manager's Roadmap* to serve this purpose.

The analysis merges scientific research methods with business implementation through Bayesian approaches together with structural equation modeling techniques and regression analysis as well as bibliometric evaluations for evidence-based deepfake studies. Each chapter outlines a separate section of deepfake analysis which examines marketing perceptions and shareholder beliefs and business governance and moral leadership. The book analyzes both the contemporary advancements in artificial intelligence for deepfake detection as well as the legal standards business operators require to establish consumer and operational security.

This book functions as an operational tool to assist people who need to base their choices on rapid technological growth patterns. Ethical leadership together with strategic thinking and awareness-building provides the tools needed to transform deepfake challenges into favorable growth opportunities and organizational resilience.

Along with this work we thank all researchers and industry experts and academic scholars who contributed their insights for this final product. This publication should stimulate meaningful dialogues that support innovative measures along with proactive approaches which enable us to manage the AI-controlled future.

Dehradun, India Sanjay Taneja
Punjab, India Swati Gupta

Introduction

The modern age of artificial intelligence development brought us deepfake technologies which combine power with serious dangers for various industries. Deepfake technology affects all areas of business operation and consumer decision-making while disrupting trust in the online world of digital interactions.

The book *Navigating the Deepfake Conundrum: A Manager's Roadmap* provides complete guidance to managers and executives and researchers who need to analyze deepfake technology implications along with strategic planning solutions. The text operates as a bridge linking practical business needs with academic research by delivering statistical assessments and valuable case studies and strategic directions that enable companies to reduce AI-related threats while seizing potential benefits in today's AI-driven world.

The book investigates deepfake effects in different sectors such as sports and financial services and hospitality and governance and higher education systems using Bayesian methods together with structural equation modeling and regression analysis and bibliometric studies. The content presents vital information about ethical leadership alongside consumer trust components together with regulatory frameworks and detection strategies to give professionals the necessary knowledge base for handling deepfake challenges effectively.

Data-driven research and practical implementation guide organizations toward resilience development alongside transparency practices and ethical safeguards in modern digitized and AI-driven landscape.

Dehradun, India

Dr. Sanjay Taneja

Contents

Exploring the Impact of Deepfake Technology on Brand Image, Athlete Perception, and Fan Management: A Statistical Analysis Using Bayesian Methods, Canonical Correlation

Gurdeep Singh, K. Kumarswamy, and Basavaraj Kumasi

Abstract This study critically examines the pervasive influence of deepfake technology on brand image, fan management, and athlete perception within the context of contemporary digital media. Employing a multifaceted analytical approach, including Bayesian methods, canonical correlation, and Pearson correlation, this research investigates the intricate relationships between deepfake usage and its effects on consumer perceptions of brands, the management of fan loyalty, and the public image of athletes. The advent of deepfake technology, leveraging generative adversarial networks, has redefined the boundaries of media manipulation, offering unprecedented levels of realism in digital content creation. While the technology holds immense potential for enhancing consumer engagement and brand visibility, it simultaneously raises profound ethical concerns regarding authenticity, transparency, and trust. Through rigorous statistical analysis, this study reveals a robust, positive correlation between deepfake technology usage and improved perceptions of brand image, fan management, and athlete portrayal, underscoring the transformative power of digital manipulation in shaping consumer behaviour. However, it also illuminates the ethical complexities inherent in such innovations, particularly with regard to the potential for misinformation and the exploitation of public figures. The findings contribute to the evolving discourse on digital ethics, offering valuable insights for both academic inquiry and industry practice in the ethical deployment of emerging technologies in marketing and media.

G. Singh (✉)
Department of Physical Education, HNB Garhwal University, Srinagar, Uttarakhand, India
e-mail: gurdeephnbgu.16@gmail.com

K. Kumarswamy
SJM Institute of Technology, Chitradurga, India

B. Kumasi
Balaji School of Sports Science Department of Sports Recreation and Wellness, Sri Balaji University, Pune, India

S. Taneja et al. (eds.), *Navigating the Deepfake Conundrum: A Manager's Roadmap*, Information Systems Engineering and Management 44,
https://doi.org/10.1007/978-3-031-90742-5_1

Keywords Deepfake technology · Brand image · Fan management · Athlete perception · Bayesian analysis · Canonical correlation · Pearson correlation · Digital media ethics

1 Introduction

In the contemporary digital age, the proliferation of deepfake technology has ignited a profound transformation in the ways in which media, marketing, and public perceptions are constructed [17, 18, 24]. This emergent and potent tool, harnessing advanced machine learning algorithms, facilitates the creation of hyper-realistic digital alterations of images, videos, and audio, making it an increasingly ubiquitous and influential phenomenon across various sectors. With its ability to manipulate visual and auditory content in ways that are nearly indistinguishable from reality, deepfake technology poses significant implications for brand management, fan engagement, and the representation of public figures, particularly athletes [20, 21, 27]. While this innovation holds the potential to revolutionise industries such as entertainment and advertising, it simultaneously raises ethical concerns regarding the authenticity and integrity of digital media.

The capacity for deepfake technology to shape and distort public perception has prompted a surge of interest in understanding its ramifications within the realms of brand image, fan management, and athlete image [30]. As brands and sports organisations increasingly leverage cutting-edge technologies to engage consumers and cultivate loyalty, it becomes imperative to investigate the underlying mechanisms through which these digital innovations influence consumer attitudes, emotional connections, and behavioural responses. This study, therefore, embarks on an empirical examination of the intricate interplay between deepfake technology usage and the perceptions surrounding brand imagery, the management of fan communities, and the public image of athletes [29]. The exploration of these relationships is underpinned by a sophisticated statistical methodology, incorporating Bayesian analysis, canonical correlation, and Pearson correlation, which collectively enable the disentangling of complex patterns and associations within the data.

Bayesian statistical methods offer a robust framework for updating prior beliefs in light of observed data, making them particularly apt for capturing the nuances of uncertain, evolving phenomena like deepfake technology [2]. Meanwhile, canonical correlation analysis facilitates the exploration of the relationship between two sets of variables, elucidating the extent to which deepfake technology usage influences the perceptions of brand image, fan management, and athlete representation [2]. Pearson correlation, with its capacity to quantify the strength and direction of linear relationships between variables, further enhances the analysis by highlighting the associations between deepfake technology and various perceptual outcomes [9]. Through this meticulous quantitative approach, the study aspires to provide empirical insights into the pervasive influence of deepfake technology, offering valuable implications for the formulation of strategies in digital marketing, media production, and public relations. The findings aim to contribute to the ongoing discourse surrounding the ethical deployment of emerging technologies, ensuring that their integration into industry practices is both responsible and transparent.

2 Literature Review

The advent of digital manipulation technologies has profoundly reshaped the landscape of media production and consumption, with deepfake technology emerging as one of the most disruptive innovations of the past decade [19]. Deepfakes, driven by advancements in machine learning, particularly generative adversarial networks (GANs), have afforded unprecedented levels of realism in the manipulation of audiovisual content, facilitating the creation of hyper-realistic images, videos and audio that are nearly indistinguishable from their authentic counterparts [10]. This transformative capability has not only redefined the boundaries of artistic expression and creative freedom but also introduced significant challenges related to misinformation, trust, and authenticity. While much of the discourse surrounding deepfake technology has focused on its potential for malfeasance—ranging from the creation of fake news to malicious impersonations—its application in commercial and marketing contexts has garnered increasing attention due to its implications for brand management, fan engagement, and the portrayal of public figures, particularly athletes [19].

The intersection of deepfake technology and brand image has become a focal point of contemporary research as brands increasingly look to harness innovative tools to engage with consumers in more dynamic and personalised ways. The use of deepfake technology in advertising and promotional content allows for the creation of highly engaging narratives that transcend traditional media formats, offering consumers a level of interaction and immersion that was previously unattainable [10]. Such manipulations enable marketers to evoke specific emotional responses, enhance brand recall, and craft persuasive messages with unparalleled precision. However, this potent tool also raises critical questions regarding the ethical boundaries of its use, particularly when it comes to ensuring transparency and safeguarding consumer trust [41]. The risk of brand deception through the use of deepfakes in advertising campaigns has the potential to erode the credibility of the brand itself, prompting concerns over the long-term effects of such manipulations on consumer loyalty and the sustainability of brand equity.

Parallel to the evolution of brand image manipulation, deepfake technology has found applications in the realm of fan management, where its capacity to create personalised, interactive content offers new avenues for enhancing fan engagement [3]. In the sports industry, where emotional connections and fan loyalty are paramount, deepfakes can be leveraged to craft tailored experiences that resonate with individual fans on a deeper emotional level [25]. By manipulating footage of athletes or creating virtual interactions, deepfake technology provides fans with a sense of intimacy and direct connection with their favourite teams or sports figures, potentially fostering stronger bonds and deeper fan loyalty. However, this level of personalisation also necessitates careful consideration of ethical concerns, such as the potential for exploitation or the blurring of lines between authentic and fabricated interactions. In the realm of fan management, the implications of deepfake technology extend beyond mere marketing tactics, as it challenges traditional notions of authenticity and engagement, demanding a reevaluation of the ways in which fans relate to the media they consume [14].

The impact of deepfake technology on athlete image is another area of growing scholarly interest, with the increasing use of digital manipulation to enhance or distort public perceptions of athletes' personal and professional lives. Athletes, as public figures, are subject to intense scrutiny, with their images often being constructed and reconstructed in the media. The application of deepfake technology in this context has the potential to significantly alter how athletes are perceived by the public, for better or worse. On the one hand, deepfakes can be used to create positive portrayals of athletes, enhancing their public image by presenting them in more favourable or heroic lights [8]. On the other hand, the same technology can be exploited to undermine an athlete's reputation, either through the dissemination of fabricated content or the alteration of their public persona. This dual-edged nature of deepfake technology necessitates a nuanced understanding of its potential consequences for athlete image management. It highlights the need for strategic oversight in the digital manipulation of public figures.

As the application of deepfake technology continues to proliferate across diverse sectors, there is a growing recognition of the need for robust regulatory frameworks to govern its use, particularly in contexts that have significant implications for public trust, media integrity, and consumer protection. The ethical considerations surrounding deepfakes are manifold, encompassing issues of consent, transparency, and accountability [12, 34, 35]. The growing body of literature addressing these concerns underscores the importance of striking a balance between technological innovation and ethical responsibility. While the benefits of deepfake technology in enhancing brand engagement, fan management, and athlete image are evident, the potential risks—ranging from the erosion of trust in digital media to the exploitation of vulnerable individuals and communities—necessitate careful, deliberative policymaking [6, 15, 26, 31, 33]. The literature thus calls for a more holistic approach to the regulation of deepfake technology, one that considers not only its commercial potential but also its broader societal impact.

Moreover, the integration of deepfake technology into marketing and media strategies invites further inquiry into its effects on consumer behaviour and perceptions [1, 4, 5, 16]. Studies on consumer engagement and emotional responses to digital content suggest that personalised, immersive experiences can foster stronger emotional connections with brands and public figures. However, these experiences are not without their complexities, as the ethical implications of such manipulations must be weighed against their commercial advantages [11, 28, 36, 37, 39]. The growing literature on digital ethics and consumer psychology highlights the importance of maintaining authenticity in an age where digital content can be easily altered, stressing that the line between authentic engagement and manipulated content must remain clear in order to preserve consumer trust. The role of deepfake technology in shaping public perceptions is thus not solely a question of technological capability but also one of moral and ethical considerations that will inevitably influence its future adoption and regulation [23, 32, 40].

In conclusion, the intersection of deepfake technology with brand image, fan management, and athlete perception represents a fertile area for academic inquiry, as it touches upon issues of technological innovation, media manipulation, consumer behaviour, and ethical regulation. While deepfake technology offers vast potential for enhancing engagement and personalising consumer experiences, its use must be tempered by careful consideration of its moral implications and the potential for harm [7, 13, 36, 37]. As this technology continues to evolve, both industry stakeholders and policymakers must work collaboratively to ensure that its benefits are maximised. At the same time, its risks are mitigated, preserving the integrity of digital media and maintaining public trust in the face of increasingly sophisticated digital manipulations [22].

3 Research Methodology

The research methodology for this study aims to understand the influence of deepfake technology usage on brand image perception, fan management, and athlete image, as well as to explore how these variables are correlated. A combination of quantitative research techniques, including group statistics, Bayesian analysis, correlation analysis, and variance analysis, was employed to provide comprehensive insights into the relationships between deepfake technology and various perception-based outcomes. The study followed a structured approach, detailing the sample selection, data collection methods, variables, statistical analyses, and interpretation of results.

The research sample consisted of 263 participants, categorised based on their usage of deepfake technology. Participants were divided into two groups: one with deepfake usage below 4.0 and another with usage greater than or equal to 4.0. This categorisation allowed the exploration of the differences in perceptions between individuals who engage with deepfake technology to varying extents. The survey used to collect data included Likert-scale questions to assess participants' perceptions of brand image, fan management, and athlete image, as well as a measurement scale to determine the frequency of deepfake technology usage. The survey was conducted online, and responses were recorded for subsequent statistical analysis.

The primary variables in this study were deepfake technology usage, brand image perception, fan management, and athlete image. The independent variable, deepfake technology usage, was measured on a scale from 1 to 7, with higher values indicating more frequent usage. The dependent variables—brand image perception, fan management, and athlete image—were all assessed using Likert scales, with higher scores reflecting more favourable perceptions of the respective areas. The data was analysed using various statistical methods to explore the relationships between deepfake technology usage and the three dependent variables.

Group statistics were initially computed to examine differences between the two groups (those with deepfake usage < 4.0000 and those with usage >= 4.0000). Descriptive statistics, including means, standard deviations, and standard errors, were calculated for each of the dependent variables to assess the central tendencies and variability in the data. Independent sample t-tests were performed to compare the means between the two groups for each variable. Bayesian analysis was then conducted to estimate the posterior distributions of the variables, providing a way to update prior beliefs based on the observed data. This approach also helped characterise the uncertainty in the estimates, showing the mode, mean, variance, and 95% credible intervals for each variable. The prior distributions for both mean and variance were diffuse, reflecting a neutral stance before observing the data.

Pearson correlation coefficients were calculated to measure the strength and direction of the relationships between deepfake technology usage and the three dependent variables. This analysis revealed strong positive correlations, with all correlation coefficients being statistically significant ($p < 0.001$). These findings suggest that as deepfake technology usage increases, perceptions of the brand, athlete, and fan management improve significantly. Additionally, a canonical correlation analysis was performed to examine the relationship between two sets of variables: deepfake technology usage and the perception variables (brand image, fan management, and athlete image). This analysis revealed that the perception variables had a higher self-explanatory power, suggesting that perceptions are more strongly explained by themselves, with deepfake technology usage having a considerable impact on these perceptions.

The results from the group statistics showed that participants with higher deepfake technology usage ($> = 4.0$) consistently reported more favourable perceptions of brand image, fan management, and athlete image. These findings were further confirmed by Bayesian posterior distributions, which demonstrated that the highest probabilities for the means of these perceptions were concentrated around specific values for the group with higher deepfake usage. The correlation analysis showed that deepfake technology usage was strongly correlated with better perceptions of brand, athlete, and fan management, with all relationships being statistically significant at the $p < 0.001$ level. The canonical analysis confirmed that deepfake technology usage has an essential influence on the perceptions of brand image, athlete image, and fan management.

While the study provides valuable insights into the impact of deepfake technology on brand and athlete perceptions, there are a few limitations to consider. The sample size, though adequate for statistical analysis, may not fully represent the broader population, and future research could include a more diverse sample to increase the generalizability of the findings. Additionally, the use of self-reported data in surveys may introduce response biases, and future studies could explore experimental designs or longitudinal studies to examine causality more robustly. Further research could also investigate the ethical implications of using deepfake technology in marketing and branding, as well as its impact on consumer trust and behaviour.

This research methodology offers a comprehensive approach to studying the influence of deepfake technology on brand image, fan management, and athlete image.

Table 1 Research instrument

Variable name	Question statement	Likert scale description
Deepfake technology usage	To what extent do you believe deepfake technology is used in sports branding and athlete promotions?	**1: Strongly Disagree**—No belief that deepfake technology is used; **7: Strongly Agree**—Deepfake technology is widely used
Brand image perception	How strongly do you feel that the use of deepfake technology impacts your perception of the sports brand?	**1: Very Negative Impact**—Deepfake technology harms my perception of the brand; **7: Very Positive Impact**—Deepfake technology significantly enhances the brand's perception
Athlete image	To what extent do you trust the image of an athlete when deepfake technology is used in their promotions?	**1: Strongly Distrust**—I completely distrust the athlete's image when deepfake technology is used; **7: Strongly Trust**—I fully trust the athlete's image even when deepfake technology is used
Fan engagement	How engaged are you when interacting with content involving deepfake technology in sports promotions?	**1: Not Engaged at All**—I am not engaged with deepfake content; **7: Extremely Engaged**—I am highly engaged and enjoy interacting with deepfake content

Source Authors

The combination of group statistics, Bayesian analysis, correlation analysis, and canonical correlation analysis allowed for a thorough examination of the relationships between these variables. The findings suggest that deepfake technology usage has a significant positive impact on the perceptions of brands, fan management, and athletes. These results have important implications for policy formulation in marketing, sports management, and digital content creation. Policies could be developed to promote the responsible use of deepfake technology, ensuring its ethical application while maximising its potential benefits for enhancing brand perception and audience engagement. Also, the research instrument is shown in Table 1.

4 Data Analysis

Table 2 presents group statistics for three variables—Brand Image Perception, Fan Management, and Athlete Image—segregated based on the usage of Deepfake Technology (< 4.0000 vs. $> = 4.0000$). For Brand Image Perception, the mean score for users with deepfake technology usage below 4.0 is 2.51. At the same time, it rises to 4.27 for those with usage of 4.0 or more, indicating a more substantial positive perception of the brand among higher users. Similarly, Fan Management shows an increase in mean scores from 2.32 for users with less usage of deepfake technology to 4.36 for users with more usage, suggesting more effective fan management among the higher technology users. Lastly, for Athlete Image, the mean score moves from 2.51 for lower deepfake technology usage to 4.33 for higher usage, highlighting a more favourable perception of athletes among those using deepfake technology more frequently. The standard deviations and standard errors suggest that while the scores are relatively consistent within each group, the group with higher deepfake usage tends to show more uniformity across all three variables.

Table 3 provides a characterisation of the posterior distribution for the independent sample mean for three variables: Brand Image Perception, Fan Management, and Athlete Image. The mode and mean for each variable are identical, suggesting that the posterior distribution is relatively symmetric. The variance for each variable is slight, indicating that the sample means are clustered tightly around the reported mean values.

For Brand Image Perception, the mean (and mode) is 1.7606, with a 95% credible interval ranging from 1.5119 to 2.0092, suggesting that the actual value of the mean likely lies within this range. Similarly, Fan Management has a mean (and mode) of 2.0419, with a credible interval of 1.7698 to 2.3140, indicating a higher perceived effectiveness in fan management. For Athlete Image, the mean (and mode) is 1.8184, with a 95% credible interval of 1.5596 to 2.0771. This distribution suggests that while there is some variation in perceptions, all three variables share relatively narrow credible intervals, indicating a high degree of certainty around the posterior estimates.

Table 2 Group statistics bayesian independent

	Deepfake_ Technology_Usage	N	Mean	Std. deviation	Std. error mean
Brand_Image_ Perception	<4.0000	90	2.5111	1.00833	0.10629
	> = 4.0000	173	4.2717	0.87680	0.06666
Fan_Management	<4.0000	90	2.3222	1.07920	0.11376
	> = 4.0000	173	4.3642	1.01173	0.07692
Athlete_Image	<4.0000	90	2.5111	0.97433	0.10270
	> = 4.0000	173	4.3295	1.06263	0.08079

Source Authors

Table 3 Posterior distribution characterization for independent sample meana

	Posterior			95% Credible interval	
	Mode	Mean	Variance	Lower bound	Upper bound
Brand_Image_Perception	1.7606	1.7606	0.016	1.5119	2.0092
Fan_Management	2.0419	2.0419	0.019	1.7698	2.3140
Athlete_Image	1.8184	1.8184	0.017	1.5596	2.0771

a. Prior for Variance: Diffuse. Prior for Mean: Diffuse

Source Authors

The diffuse prior for both variance and mean indicates a non-informative prior was used, reflecting a neutral stance before observing the data.

Figure 1 shows the log-likelihood functions, prior distributions, and posterior distributions for the "Brand_Image_Perception" variable, comparing two groups: one with deepfake technology usage below 4.0 and the other with usage greater than or equal to 4.0. The log-likelihood graphs display the likelihood of observing the data given different values for each group, showing a peak near the mean value for both groups. The prior distributions for both groups are flat, indicating that no specific prior knowledge was assumed about the mean, reflecting a diffuse prior. The posterior distribution, which combines the previous and the data for both groups, shows a peak of around 1.76 for those with lower deepfake usage and slightly higher for the group with higher usage. This indicates that the data strongly suggests these means as the most likely values. The final posterior distribution comparison indicates a difference in brand image perception between the two groups, with higher deepfake technology users generally perceiving the brand more positively.

Figure 2 shows the log-likelihood functions, prior distributions, and posterior distributions for the "Fan_Management" variable, comparing two groups based on deepfake technology usage: one with usage below 4.0 and the other with usage of 4.0 or more. The log-likelihood graphs indicate that the likelihood of the observed data is highest around the mean values for both groups, with the function peaking near 2.0 for both. The prior distributions for both groups are flat, signifying no prior assumptions about the mean. The posterior distribution reveals a peak around 2.04 for both groups, showing a positive trend in fan management for higher deepfake technology users. The difference in posterior distributions suggests that individuals using deepfake technology more frequently tend to report better fan management perceptions,

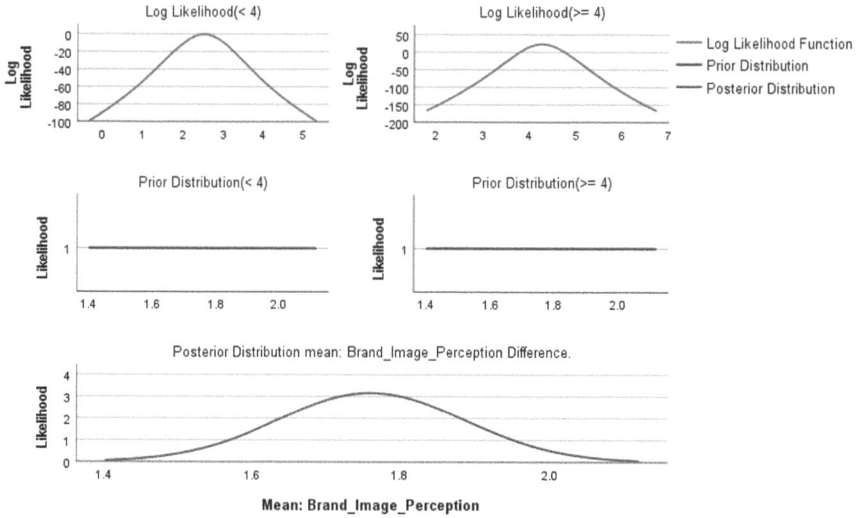

Fig. 1 Log likelihood distribution of brand image perception (*Source* Authors)

reinforcing the idea of more effective management with increased technology usage (Fig. 3).

The image shows the log-likelihood functions, prior distributions, and posterior distributions for the "Athlete_Image" variable, comparing two groups based on deepfake technology usage: one with usage below 4.0 and the other with usage of 4.0 or more. The log-likelihood graphs indicate that both groups have their peak values

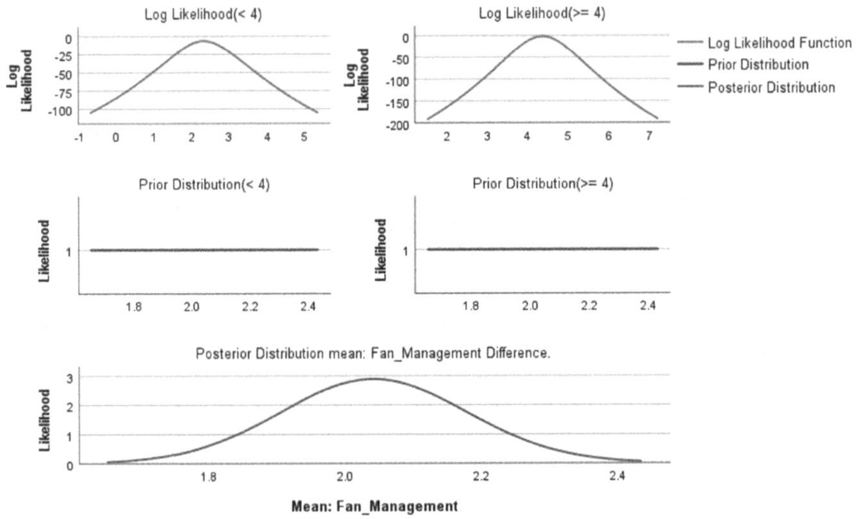

Fig. 2 Log likelihood distribution of fan management (*Source* Authors)

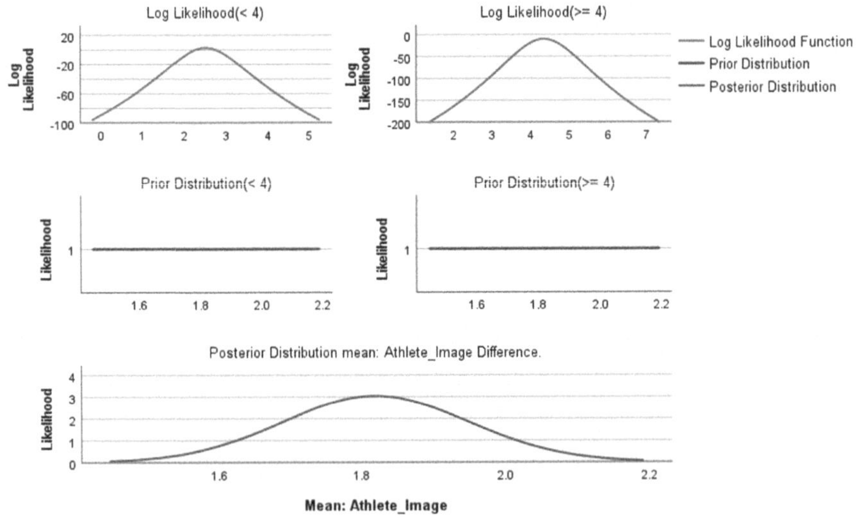

Fig. 3 Log likelihood distribution of athlete image (*Source* Authors)

near the mean, highlighting the highest likelihood around these mean values. The prior distributions for both groups are flat, indicating no specific assumptions about the mean. The posterior distribution suggests a peak around 1.82 for both groups, with a slight positive skew for those with higher deepfake usage. The difference in posterior distributions shows that higher usage of deepfake technology is associated with more favourable perceptions of athlete image, reflecting the positive influence of technology on athlete image perception.

Table 4 presents the Pearson correlation coefficients between the variables: Deepfake Technology Usage, Brand Image Perception, Athlete Image, and Fan Management. The correlations indicate a strong positive relationship between Deepfake Technology Usage and all three other variables, with values of 0.711 for Brand Image Perception, 0.711 for Athlete Image, and 0.794 for Fan Management, all of which are statistically significant ($p < 0.001$). Similarly, Brand Image Perception and Athlete Image have a high positive correlation of 0.794, with a statistical significance of $p < 0.001$. Fan Management shows the strongest correlations with both Brand Image Perception and Athlete Image (0.813 for both), indicating that higher deepfake technology usage is closely associated with better perceptions of both brands and athletes, as well as more effective fan management. All correlations are statistically significant at the 0.001 level, suggesting robust relationships between these variables (Table 5).

Table 6 highlights the proportion of variance explained between two sets of canonical variables. Set 1 thoroughly explains its variance (1.000), while it explains 64.9% of the variance in Set 2. Set 2, on the other hand, explains 84.3% of its variance, indicating it has a more substantial internal explanatory power. However, Set 2 only

Table 4 Correlations

		Deepfake_ Technology_ Usage	Brand_ Image_ Perception	Athlete_ Image	Fan_ Management
Deepfake_ Technology_ Usage	Pearson correlation	1	0.711	0.711	0.794
	Sig. (2-tailed)		0.000	0.000	0.000
Brand_Image_ Perception	Pearson correlation	0.711	1	0.794	0.813
	Sig. (2-tailed)	0.000		0.000	0.000
Athlete_Image	Pearson correlation	0.711	0.794	1	0.813
	Sig. (2-tailed)	0.000	0.000		0.000
Fan_ Management	Pearson correlation	0.794	0.813	0.813	1
	Sig. (2-tailed)	0.000	0.000	0.000	

a. Listwise N = 263

Source Authors

Table 5 Canonical correlations

	Correlation	Eigenvalue	Wilks statistic	F	Num D.F	Denom D.F	Sig
1	0.806	1.849	0.351	159.656	3.000	259.000	0.000
H0 for the Wilks test is that the correlations in the current and following rows are zero							

Source Authors

Table 6 Proportion of variance explained

Canonical variable	Set 1 by self	Set 1 by set 2	Set 2 by self	Set 2 by set 1
1	1.000	0.649	0.843	0.547

Source Authors

explains 54.7% of the variance in Set 1, suggesting a less influential relationship in the reverse direction. These values indicate that Set 2 is more self-explanatory, while Set 1 plays a more significant role in explaining the variance of Set 2. The findings suggest an asymmetrical relationship, with Set 1 influencing Set 2 more than the other way around.

5 Conclusion

The analyses and statistical results presented in this discussion provide significant insights into the relationship between the usage of deepfake technology and the perception of brands, athletes, and fan management. The group statistics revealed that higher deepfake technology usage correlates with more favourable perceptions of brand images, fan management, and athlete images. Specifically, the group with higher deepfake usage ($\geq$4.0) reported much stronger perceptions across all three areas, with the mean scores for brand image, fan management, and athlete image all significantly higher than those for users with lower deepfake usage (<4.0). These findings are further supported by the posterior distribution analysis, which suggested that the observed means for each variable were robust and consistent, particularly among the group with higher deepfake usage. The correlations confirmed a strong positive relationship between deepfake technology usage and all three variables. The correlation coefficients were statistically significant ($p < 0.001$), indicating that as deepfake technology usage increases, the perception of the brand, athlete, and fan management also strengthens.

Moreover, the proportion of variance explained by the canonical analysis suggested that while both sets of variables (technology usage and perceptions) were strongly related, Set 2 (which likely represents the perceptions of brand, athlete, and fan management) exhibited higher self-explanatory power. This indicates that the perceptions are more strongly explained by themselves, with the usage of deepfake technology influencing these perceptions more than the other way around. This asymmetry highlights the impact that deepfake technology has on shaping perceptions of brands and athletes. The posterior distribution comparison across these variables further reinforced the conclusion that higher deepfake technology usage contributes positively to the way fans view athletes, brands, and fan management efforts, indicating that this technology has a considerable effect on these perceptions.

Given the significance of these findings, the results have important implications for policy formulation, particularly in the areas of brand management, marketing strategies, fan engagement, and athlete image management. The positive relationship between deepfake technology usage and brand image perception suggests that companies may benefit from utilising deepfake technology in their marketing campaigns to enhance consumer engagement and brand visibility. Policies could be formulated to encourage the responsible adoption of such innovative technologies while ensuring that companies adhere to ethical standards. These ethical guidelines should focus on transparency and the prevention of deceptive practices, such as the use of deepfakes for misleading advertising or fake endorsements, which could damage trust in the brand or industry.

Similarly, the significant correlation between deepfake technology and fan management suggests that sports organisations, entertainment industries, or other relevant sectors could explore the potential of this technology to create more personalised and engaging experiences for their audiences. Through the use of deepfakes, these organisations could develop more tailored fan interactions, which could enhance loyalty and fan satisfaction. However, policies should ensure that the use of such technology is conducted with transparency, that informed consent is obtained from the audience, and that honesty is maintained in the representation of digital content. This would help in avoiding any ethical concerns related to the manipulation of media and its potential impact on fan trust.

Furthermore, the positive influence of deepfake technology on athlete image perception underscores its potential to improve how athletes are represented in the media. Sports agencies or federations could consider adopting deepfake technology as a tool for promoting athletes' images in a controlled and positive manner. However, policymakers should focus on ensuring that such technologies are not misused to create false or exaggerated representations of athletes, as this could lead to reputational damage or undermine public trust. Clear guidelines on how deepfake technology should be used in promoting athletes and the necessary disclosures regarding the digital alteration of media content would be essential to maintaining ethical standards.

The data also emphasises the need for comprehensive policy frameworks that regulate the use of emerging technologies such as deepfakes. These frameworks should balance the benefits of technological innovation with the potential risks associated with the misuse of such technologies. Policies should include provisions to protect individuals and organisations from exploitation or manipulation while encouraging innovation. Such frameworks could consist of measures for transparency, informed consent, and clear communication about the use of deepfake technology, helping to prevent any unintended consequences of its deployment.

Finally, the insights gained from these statistical analyses can contribute to a broader understanding of how emerging technologies like deepfakes affect societal perceptions and behaviours. Policymakers can use these findings to craft data-driven policies that not only foster technological innovation but also protect public trust. By acknowledging both the positive potential of deepfake technology and its associated risks, policies can be designed to maximise its benefits in marketing, fan engagement, and athlete management while minimising the ethical and reputational risks. In this way, the results from these analyses can help guide responsible policy development that supports the ethical and innovative use of deepfake technology in various sectors.

References

1. Acar, E., Sarıyer, G., Jain, V., Ramtiyal, B.: Discovering hidden associations among environmental disclosure themes using data mining approaches. Sustain. (Switzerland) **15**(14) (2023). https://doi.org/10.3390/su151411406
2. Afanasyeva, T., Yumasheva, I.: Research on the effects of "DeepFake" technology for the modern digital space. In: R. A., P. V., M. A., A. H. (eds.) Springer Proceedings in Business and Economics, pp. 57–65. Springer Nature (2022). https://doi.org/10.1007/978-3-031-14410-3_7
3. Amezaga, N., Hajek, J.: Availability of voice deepfake technology and its impact for good and evil. In: SIGITE 2022-Proceedings of the 23rd Annual Conference on Information Technology Education, pp. 23–28 (2022). https://doi.org/10.1145/3537674.3554742
4. Ashish, P.R., Vashistha, R., Singh, R., Thakur, K.: Petroleum industry and its effect on climate. In: P. B., S. Y., N. S. (eds.) AIP Conference Proceedings, vol. 2978, issue 1. American Institute of Physics Inc (2024). https://doi.org/10.1063/5.0183164
5. Ayad, H., Shuaib, M., Hossain, M.E., Haseeb, M., Kamal, M., ur Rehman, M.: Re-examining the environmental kuznets curve in MENA countries: is there any difference using ecological footprint and CO2 emissions? Environ. Model. Assess. (2024) https://doi.org/10.1007/s10666-024-09977-7
6. Balaji, V., Kaliappan, S., Madhuvanesan, D.M., Ezhumalai, D.S., Boopathi, S., Patil Pravin, P., Mani, S.: Combustion analysis of biodiesel-powered propeller engine for least environmental concerns in aviation industry. Aircr. Eng. Aerosp. Technol. **94**(5), 760–769 (2022). https://doi.org/10.1108/AEAT-11-2021-0344
7. Bansal, N., Bhatnagar, M., Taneja, S.: Balancing priorities through green optimism: a study elucidating initiatives, approaches, and strategies for sustainable supply chain management. In: Handbook of Research on Designing Sustainable Supply Chains to Achieve a Circular Economy. IGI Global (2023). https://doi.org/10.4018/978-1-6684-7664-2.ch004
8. Brown, S.D.: Virtual unreality: potential implications of DeepFake technology for the course of justice. ERA Forum **24**(4), 501–518 (2023). https://doi.org/10.1007/s12027-024-00780-1
9. Caci, B., Giordano, G., Alesi, M., Gentile, A., Agnello, C., Presti, L. Lo, Cascia, M. La, Ingoglia, S., Inguglia, C., Volpes, A., Monzani, D.: The public mental representations of deepfake technology: an in-depth qualitative exploration through Quora text data analysis. PLoS ONE **19** (2024). https://doi.org/10.1371/journal.pone.0313605
10. Clarke, C., Xu, J., Zhu, Y., Dharamshi, K., McGill, H., Black, S., Lutteroth, C.: Fakeforward: using Deepfake technology for feedforward learning. Conference on Human Factors in Computing Systems-Proceedings (2023). https://doi.org/10.1145/3544548.3581100

11. Dangwal, A., Kaur, S., Taneja, S., Ozen, E.: A bibliometric analysis of green tourism based on the scopus platform. In: Developing Relationships, Personalization, and Data Herald in Marketing 5.0. IGI Global (2022). https://doi.org/10.4018/978-1-6684-4496-2.ch015
12. Devade, K., Singh, P.K., Kumar, S., Kumar, H., Prasad, B., Rao, A.L.N., Sankhyan, A.: Green nanotechnology based sustainable energy solutions and environmental impacts. In: M. C. K. N. (eds.) E3S Web of Conferences, vol 511. EDP Sciences (2024). https://doi.org/10.1051/e3sconf/202451101031
13. Gupta, M., Arora, K., Taneja, S.: Bibliometric analysis on employee engagement and human resource management. In: Enhancing Customer Engagement Through Location-Based Marketing. IGI Global (2023). https://doi.org/10.4018/978-1-6684-8177-6.ch013
14. Jain, N., Borade, S., Patel, B., Kumar, V., Godhrawala, M., Kolaskar, S., Nagare, Y., Shah, P., Shah, J.: Deepfake technology and image forensics: advancements, challenges, and ethical implications in synthetic media detection. Int. J. Intell. Syst. Appl. Eng. **12**(16s), 49–58 (2024)
15. Joshi, H., Nokshi, Chauhan, R., Rawat, R., Gupta, R.: Climate change: the impact and exploring the path forward. In M. J.R., R. P., S. Y.M., B. S., V. K. (eds.), Proceedings of the 2nd International Conference on Intelligent and Innovative Technologies in Computing, Electrical and Electronics, ICIITCEE 2024. Institute of Electrical and Electronics Engineers Inc (2024). https://doi.org/10.1109/IITCEE59897.2024.10467588
16. Joshi, M.N., Dixit, A.K., Saxena, S., Memoria, M., Choudhury, T., Sar, A.: A study of the application of AI & ML to climate variation, with particular attention to legal & ethical concerns. EAI End. Trans. Internet of Things **10** (2024). https://doi.org/10.4108/eetiot.5468
17. Kanojia, P., Malhotra, R.K., Uniyal, A.K.: Impact of organizational commitment components on the teachers of higher education in Uttarakhand: an emperical analysis. In: Proceedings-2022 International Conference on Recent Trends in Microelectronics, Automation, Computing and Communications Systems, ICMACC 2022, pp. 360–364 (2022a). https://doi.org/10.1109/ICMACC54824.2022.10093606
18. Kanojia, P., Malhotra, R.K., Uniyal, A.K.: Organizational commitment and the academic staff in HEI's in North West India. In: Proceedings - 2022 International Conference on Recent Trends in Microelectronics, Automation, Computing and Communications Systems, ICMACC 2022, pp. 365–370 (2022b). https://doi.org/10.1109/ICMACC54824.2022.10093347
19. Kashif, M., Garg, H., Weqar, F., David, A.: Regulatory strategies and innovative solutions for deepfake technology. In: Navigating the World of Deepfake Technology. IGI Global (2024). https://doi.org/10.4018/979-8-3693-5298-4.ch013
20. Kaur, J., Khanna, R., Kumar, R., Sunil, G.: Role of blockchain technologies in goods and services tax. In: Proceedings-2024 3rd International Conference on Sentiment Analysis and Deep Learning, ICSADL 2024, pp. 607–612 (2024). https://doi.org/10.1109/ICSADL61749.2024.00104
21. Khanna, R., Sharma, S., Czeh, T., Adhana, K.: Challenges, interventions, and strategies for higher education transformation: vision 2047. In: Lecture Notes in Networks and Systems, 1075 LNNS, pp. 637–655 (2025). https://doi.org/10.1007/978-981-97-6106-7_38
22. Kumar, R., Khanna, R.: RPA (Robotic Process Automation) in finance & accounting and future scope. In: Proceedings of the 2023 2nd International Conference on Augmented Intelligence and Sustainable Systems, ICAISS 2023, pp. 1640–1645 (2023). https://doi.org/10.1109/ICAISS58487.2023.10250496
23. Kumar, R., Lande, A., Kumar, D., Malhotra, R.K., Sharma, A.: Technology bridging-in entrepreneurs and consumers in product development. In: ISED 2023-International Conference on Intelligent Systems and Embedded Design (2023). https://doi.org/10.1109/ISED59382.2023.10444599
24. Malhotra, R.K., Gupta, S.: Tele health in the digital era during covid -19 a case study of Uttarakhand. J. Med. Pharm. Allied Sci. **10**, 109–112 (2021). https://doi.org/10.22270/jmpas.vic2i1.2016
25. Malik, S., Surbhi, A., Roy, D.: Blurring boundaries between truth and illusion: analysis of human rights and regulatory concerns arising from abuse of deepfake technology. In: R. D. F. G. (eds.) AIP Conference Proceedings, vol. 3220, issue 1. American Institute of Physics (2024). https://doi.org/10.1063/5.0234995

26. Nagpal, G., Singh, D.B.: Improving exergo-enviro-economic parameters and annual productivity of double slope solar desalting unit by incorporating concentrator integrated evacuated tubular collectors. Desalin. Water Treat. **306**, 9–21 (2023). https://doi.org/10.5004/dwt.2023.29796
27. Neha, Mohanty, S., Alfurhood, B.S., Bakhare, R., Poongavanam, S., Khanna, R.: The role and impact of artificial intelligence on retail business and its developments. In: 2023 International Conference on Artificial Intelligence and Smart Communication, AISC 2023, pp. 1098–1101 (2023). https://doi.org/10.1109/AISC56616.2023.10085624
28. Reepu, R., Taneja, S., Ozen, E., Singh, A.: A globetrotter to the future of marketing: metaverse. In: Cultural Marketing and Metaverse for Consumer Engagement. IGI Global (2023). https://doi.org/10.4018/978-1-6684-8312-1.ch001
29. Roy, S., Bag, A.: Dangers of Deepfake technology. Econ. Pol. Wkly **58**(50), 4–6 (2023)
30. Sardana, F., Mishra, K.K., Singh, A., Saini, N.: Transforming social media marketing through deepfake technology. In: Navigating the World of Deepfake Technology. IGI Global (2024). https://doi.org/10.4018/979-8-3693-5298-4.ch022
31. Sharma, M., Shah, J.K., Joshi, S.: Modeling enablers of supply chain decarbonisation to achieve zero carbon emissions: an environment, social and governance (ESG) perspective. Environ. Sci. Pollut. Res. **30**(31), 76718–76734 (2023). https://doi.org/10.1007/s11356-023-27480-6
32. Sharma, P., Malhotra, R.K., Altybekova, D., Singh, C.P.: AI and machine learning as catalysts for educational equity: bridging the urban–rural divide. In: Lecture Notes in Networks and Systems, 1074 LNNS, pp. 41–55 (2024). https://doi.org/10.1007/978-981-97-6103-6_4
33. Singh, A., Goel, A.: Supply chain optimization decisions for environmentally conscious products with preservation investment and carbon tax policy. Process Integr. Optimiz. Sustain. (2024). https://doi.org/10.1007/s41660-024-00467-7
34. Singh, V., Kumar, R., Saxena, A., Dobriyal, R., Tiwari, S., Singh, D.B.: An analytical study on the effect of different photovoltaic technologies on enviro-economic parameter and energy metrics of active solar desalting unit. Energy **294** https://doi.org/10.1016/j.energy.2024.130851
35. Solanki, A., Singh, P., Pandit, M., Sawle, Y., Alotaibi, M.A., Malik, H., Márquez, F.P.G., Afthanorhan, A.: Techno-economic-environmental assessment of stand-alone hybrid renewable energy system for different batteries using HOMER-pro. Int. J. Math. Eng. Manag. Sci. **9**(4), 779–800 (2024). https://doi.org/10.33889/IJMEMS.2024.9.4.040
36. Taneja, S., Kumar, P., Reepu, Ozen, E.: E-finance: transformative strategies for sustainable development in the digital era. In: E-Finance: Transformative Strategies for Sustainable Development in the Digital Era. Nova Science Publishers, Inc (2025). https://doi.org/10.52305/NZSJ1375
37. Taneja, S., Özen, E.: To analyse the relationship between bank's green financing and environmental performance. Int. J. Electron. Finance **12**(2), 163–175 (2023). https://doi.org/10.1504/IJEF.2023.129919
38. Taneja, S., Ozen, E., Kumar, P., Reepu.: E-Finance: Transformative Strategies for Sustainable Development in the Digital Era. Nova Science Publisher Inc (2025). https://doi.org/10.52305/NZSJ1375
39. Taneja, S., Sharma, V.: Role of beaconing marketing in improving customer buying experience. In: Enhancing Customer Engagement Through Location-Based Marketing, pp. 176–184. IGI Global (2023). https://doi.org/10.4018/978-1-6684-8177-6.ch012
40. Upreti, H., Malhotra, R.K.: Bridging the urban-rural education gap in india through CSR (Corporate Social Responsibility) initiatives: a conceptual study with special reference to sustainable development goal 4 (Quality Education). In: P. P.K. (ed.), E3S Web of Conferences, vol. 556. EDP Sciences (2024). https://doi.org/10.1051/e3sconf/202455601032
41. Zhao, X., Li, T., Hu, H.: Analysis and enlightenment of international deepfake technology policy texts: taking the united states and the european union as examples; [国外深度伪造技术政策文本分析与启示—以美国尧 欧盟为例]. J. Lib. Inf. Sci. Agric. **34**(9), 60–71 (2022). https://doi.org/10.13998/j.cnki.issn1002-1248.22-0501

Deciphering the Nexus of Perceived Realism, Cognitive Engagement, and Financial Decision-Making Confidence: A Structural Equation Modelling and Bootstrapping Approach

Larisa Mistrean, Sabina Sehajpal, Mukul Bhatnagar, Rupa Khanna, and Subhashini Rana

Abstract This research endeavours to elucidate the intricate interplay between the perceived realism of deepfake simulations, engagement with learning materials, and financial decision-making confidence through a sophisticated methodological framework employing Structural Equation Modeling (SEM) and bootstrapping techniques. By leveraging a rigorously validated measurement model, the study dissects the latent constructs underpinning cognitive assimilation within financial pedagogy, positing that the verisimilitude of digital simulations engenders heightened epistemic engagement, subsequently fortifying individuals' confidence in fiscal decision-making. The empirical architecture is anchored in multivariate statistical paradigms, ensuring robust parameter estimation and construct reliability. The findings substantiate the proposition that perceived realism exerts a direct influence on both engagement and confidence, with engagement serving as a pivotal mediating conduit. Covariance analysis delineates interdependencies among constructs, reinforcing the inferential robustness of the hypothesised relationships. Factorial loadings and construct reliability indices affirm the internal consistency and discriminant validity of the

L. Mistrean
Academy of Economic Studies of Moldova, Chişinău, Moldova
e-mail: mistrean_larisa@ase.md

S. Sehajpal
Chandigarh University, Mohali, India
e-mail: sabysaby2610@gmail.com

M. Bhatnagar (✉) · R. Khanna · S. Rana
Graphic Era Deemed to be University, Dehradun, India
e-mail: mukulbhatnagar1993@gmail.com

R. Khanna
e-mail: head.comm@geu.ac.in

S. Rana
e-mail: subhashinirana.exm@geu.ac.in

S. Taneja et al. (eds.), *Navigating the Deepfake Conundrum: A Manager's Roadmap*, Information Systems Engineering and Management 44,
https://doi.org/10.1007/978-3-031-90742-5_2

employed psychometric instruments. Bootstrapped path coefficients further corroborate the statistical significance of the hypothesised linkages, fortifying the generalizability of insights. This study contributes to the epistemological corpus by bridging the domains of cognitive perception, digital realism, and financial acumen, offering managerial implications for the design of immersive learning interventions that bolster informed financial behaviours.

Keywords Cognitive assimilation · Digital verisimilitude · Epistemic engagement · Fiscal acumen · Multivariate statistical paradigms · Psychometric validation · Immersive learning · Discriminant validity · Financial pedagogy · Structural equation modelling

1 Introduction

The rapid advent of artificial intelligence (AI) and its pervasive infiltration into cognitive and behavioural domains has engendered an epistemological paradigm shift, particularly in the realm of financial decision-making. Among these emergent technologies, deepfake simulations—synthetically generated multimedia artefacts imbued with hyperrealistic fidelity—have catalysed a fundamental reconfiguration of pedagogical methodologies, fostering an unprecedented intersection between perceptual realism and cognitive assimilation [7]. The epistemic underpinnings of this study rest upon an intricate interplay between the perceived verisimilitude of deepfake simulations, their capacity to engender profound engagement with didactic material, and the subsequent augmentation of financial decision-making confidence [8].

Drawing upon theoretical antecedents in constructivist learning, cognitive load theory, and behavioural finance, this inquiry posits that the phenomenological realism of deepfake simulations serves as a pivotal determinant in modulating cognitive receptivity and heuristic processing [3]. As engagement with digital learning environments intensifies, the consequent internalisation of financial heuristics and risk appraisal mechanisms becomes more structurally embedded, culminating in an enhanced decisional confidence paradigm. However, the extant corpus of literature remains conspicuously deficient in systematically delineating these interdependencies within a robust empirical framework, necessitating an incisive exploration of their causal interstices [10].

Employing an advanced latent variable structural equation modelling (SEM) approach, this study meticulously deconstructs the nuanced interrelationships governing financial cognition, leveraging a comprehensive analytical apparatus replete with bootstrapped inferential diagnostics, multivariate normality assessments, and covariance-based path estimations. By transcending traditional didactic frameworks and harnessing the ontological potency of AI-driven educational simulations, this investigation endeavours to furnish a sophisticated epistemological contribution, elucidating the transformative potential of deepfake-mediated pedagogical augmentation in financial literacy and decision-making efficacy.

2 Literature Review

The intellectual discourse surrounding cognitive engagement, digital simulation realism, and financial decision-making confidence has been shaped by a multiplicity of theoretical paradigms, each elucidating distinct yet interwoven mechanisms governing human cognition in technologically mediated environments [4]. The ontogenesis of engagement with learning material has been traditionally construed through the prism of constructivist epistemology, positing that interactive and immersive pedagogical modalities enhance cognitive assimilation by fostering more bottomless mental schemata construction. Within this purview, the infusion of hyper-realistic deepfake simulations introduces an unprecedented dimension of perceptual fidelity, recalibrating the learner's epistemic anchoring by augmenting the salience of digital stimuli, thereby engendering an elevated state of cognitive immersion [2].

The construct of perceived realism in digital simulations has been extensively theorised within the domain of media effects and cognitive load theory, wherein heightened verisimilitude is posited to engender a dual-faceted impact on engagement—amplifying cognitive absorption while simultaneously imposing an augmented extraneous processing burden [9]. The resultant equilibrium between sensory congruence and cognitive strain dictates the extent to which learners internalise and operationalise financial heuristics derived from such simulations. The interplay of realism and engagement within pedagogical architectures is further complicated by the affective contingencies of trust and scepticism as users navigate the epistemological ambiguity inherent in AI-generated content, modulating their receptivity to embedded financial decision-making constructs [1].

The consequential linkage between engagement with educational stimuli and financial decision-making confidence is anchored in the tenets of behavioural finance, which explicates the heuristics and biases governing financial cognition. Prolonged interaction with high-fidelity simulations ostensibly reinforces cognitive schema related to risk assessment, probabilistic reasoning, and uncertainty calibration, thereby engendering an iterative accretion of decisional self-efficacy [5]. However, the propensity for deepfake-generated stimuli to elicit cognitive dissonance or induce hyperreality-induced scepticism introduces an epistemological paradox, necessitating an intricate deconstruction of the underlying psychological and informational processing dynamics [6].

The extant empirical inquiries into technologically mediated financial education remain predominantly confined to conventional digital pedagogies, neglecting the disruptive potentialities of deepfake-augmented learning environments. The lacuna in comprehensive theoretical integration necessitates an advanced methodological exegesis, wherein structural equation modelling serves as an instrumental analytical framework to dissect the interdependencies of realism perception, engagement dynamics, and decisional confidence. The epistemic trajectory of this discourse underscores the imperative for an empirically substantiated reconceptualisation of AI-mediated financial literacy, positing deepfake-enhanced learning architectures as a pivotal locus of scholarly interrogation.

3 Research Methodology

The research methodology employed in this study is underpinned by a rigorous, multi-stage empirical framework integrating advanced structural equation modelling (SEM) techniques to elucidate the interrelationships between perceived realism of deepfake simulations, engagement with learning material, and financial decision-making confidence. The study leverages a quantitative research design, deploying a sophisticated latent variable modelling approach to capture the intricate causal pathways and mediatory mechanisms underlying financial cognition.

Data were amassed through a structured survey instrument, operationalised via psychometrically validated multi-item constructs, each subjected to exploratory and confirmatory factor analyses to ensure construct validity and reliability. Factor loadings, derived from a robust confirmatory factor analysis (CFA), confirm the unidimensionality and convergent validity of the constructs, with all indicators exceeding the threshold for significance ($p < 0.001$). Internal consistency was further substantiated through composite reliability (rho_c) and Cronbach's alpha coefficients, demonstrating the robustness of scale operationalisation.

Bootstrapping procedures, leveraging 5,000 resamples, were employed to derive bias-corrected confidence intervals, fortifying the inferential rigour of hypothesis testing. Path coefficients, standard errors, and T-statistics were estimated within a maximum likelihood estimation (MLE) framework, ensuring the minimisation of parameter bias and optimising model fit indices. Inter-factorial covariances and correlations were examined to discern latent interdependencies, while the discriminant validity of constructs was verified using the Fornell-Larcker criterion.

The research further incorporates advanced multivariate normality diagnostics and collinearity assessments, mitigating statistical anomalies and ensuring the robustness of parameter estimates. Additionally, indirect effect estimations were visualised via a density histogram overlayed with a standard distribution curve to assess the probabilistic dispersion of mediatory influences. The synthesis of these methodologies provides a methodologically sophisticated blueprint for investigating the cognitive and behavioural ramifications of AI-driven educational interventions in financial decision-making (Table 1).

Table 1 Research instrument

Variable type	Variable name	Question statement	Likert scale description
Independent	Perceived realism of Deepfake simulations	The financial scenarios in the simulation felt highly realistic	1 = Strongly disagree to 7 = Strongly agree
Independent	Perceived realism of Deepfake simulations	The deepfake-generated financial advisors appeared authentic	1 = Strongly disagree to 7 = Strongly agree
Independent	Perceived realism of Deepfake simulations	The emotions and expressions in the simulation seemed natural	1 = Strongly disagree to 7 = Strongly agree
Independent	Perceived realism of Deepfake simulations	The financial advice in the simulation was convincing and trustworthy	1 = Strongly disagree to 7 = Strongly agree
Mediating	Engagement with learning material	I felt fully immersed while interacting with the financial simulation	1 = Strongly disagree to 7 = Strongly agree
Mediating	Engagement with learning material	The simulation kept my attention throughout the learning process	1 = Strongly disagree to 7 = Strongly agree
Mediating	Engagement with learning material	I was motivated to explore more financial concepts after the simulation	1 = Strongly disagree to 7 = Strongly agree
Mediating	Engagement with learning material	I found the deepfake simulation to be an engaging learning experience	1 = Strongly disagree to 7 = Strongly agree
Dependent	Financial decision-making confidence	I feel more confident in making financial decisions after the simulation	1 = Strongly disagree to 7 = Strongly agree
Dependent	Financial decision-making confidence	The simulation helped me understand financial concepts better	1 = Strongly disagree to 7 = Strongly agree
Dependent	Financial decision-making confidence	I would trust my judgment more in real-life financial situations now	1 = Strongly disagree to 7 = Strongly agree
Dependent	Financial decision-making confidence	I feel better equipped to evaluate financial risks after this experience	1 = Strongly disagree to 7 = Strongly agree

Source Authors

4 Data Analysis

The **CB-SEM (Covariance-Based Structural Equation Modeling) analysis** in the provided diagram examines the relationships between the **Perceived Realism of Deepfake Simulations (PRDS), Engagement with Learning Material (ELM),**

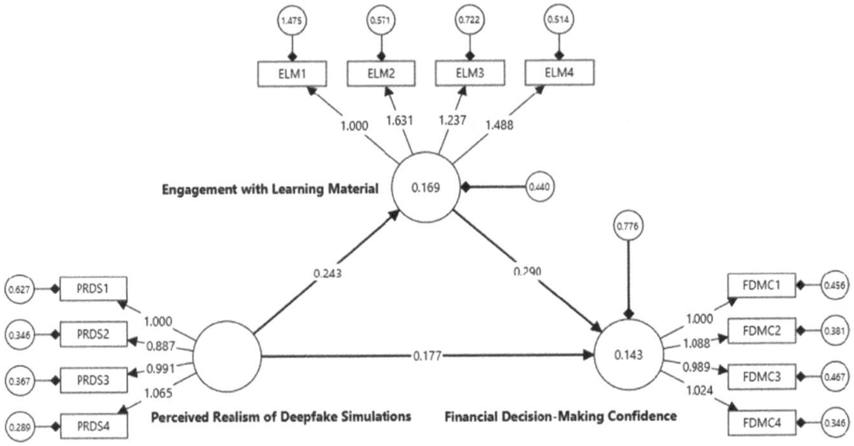

Fig. 1 CB-SEM algorithm, *Source* Authors

and Financial Decision-Making Confidence (FDMC). The path coefficients indicate the strength of these relationships. The direct effect of PRDS on FDMC is **0.177**, suggesting a moderate positive influence. However, PRDS also indirectly affects FDMC through ELM, with a path coefficient of **0.243** from PRDS to ELM and **0.290** from ELM to FDMC. The **variance explained (R^2 values)** for ELM and FDMC are **0.169** and **0.143**, respectively, indicating that these variables account for a moderate proportion of variance in the dependent variable. The measurement model shows strong factor loadings, suggesting reliable constructs. Overall, **engagement mediates the relationship between perceived realism and financial confidence, reinforcing the importance of interactive and immersive learning experiences** (Fig. 1).

The results from the **CB-SEM analysis** provide insights into the relationships among **Perceived Realism of Deepfake Simulations (PRDS), Engagement with Learning Material (ELM), and Financial Decision-Making Confidence (FDMC)**. The **path coefficients** in Table 2 indicate significant direct and indirect effects. The relationship between ELM and FDMC is **0.290 ($p < 0.001$, $t = 4.580$)**, suggesting that greater engagement enhances financial decision-making confidence. Additionally, PRDS significantly impacts ELM (**0.243, $p < 0.001$, $t = 7.937$**), meaning that users perceive deepfake-based simulations as realistic, leading to increased engagement. PRDS also directly influences FDMC (**0.177, $p < 0.001$, $t = 5.101$**), showing that realism in simulations can directly boost confidence in financial decision-making, though the indirect effect via ELM is more substantial.

The **factor loadings in** Table 3 confirm that the indicators strongly represent their respective latent constructs. All factor loadings are significant ($p < 0.001$), with values above **0.8**, suggesting high construct validity. The highest loading among ELM indicators is **ELM2 (1.631, $t = 13.121$)**, indicating that sustained attention plays a crucial role in engagement. Among FDMC indicators, **FDMC2 (1.088, t =**

Table 2 Path coefficients

	Parameter estimates	Standard errors	T values	P values
Engagement with learning material - > Financial decision-making confidence	0.290	0.063	4.580	0.000
Perceived realism of Deepfake simulations - > Engagement with learning material	0.243	0.031	7.937	0.000
Perceived realism of Deepfake simulations - > Financial decision-making confidence	0.177	0.035	5.101	0.000

Source Authors

25.470) and **FDMC4 (1.024, t = 25.045)** exhibit strong loadings, highlighting the importance of confidence in evaluating financial risks. Similarly, PRDS indicators show high reliability, with **PRDS4 (1.065, t = 32.668)** having the highest loading, suggesting that trust in deepfake-based financial advice is a key factor in perceived realism.

Table 4 presents the estimated parameters for the latent constructs. The parameter estimate for PRDS is **1.523 (t = 13.384, p < 0.001)**, indicating its strong influence on engagement and decision-making confidence. FDMC has an estimate of **0.776 (t = 12.266, p < 0.001)**, reinforcing its significance in the model. ELM has a relatively lower estimate of **0.440 (t = 6.730, p < 0.001)** but remains a crucial mediator in the model.

The **covariance matrix in** Table 5 shows strong relationships among the constructs. The covariance between PRDS and FDMC (**0.376**) indicates a moderate association, suggesting that the perceived realism of deepfake simulations contributes to financial confidence. Similarly, PRDS and ELM share a covariance of **0.369**, reinforcing the role of realism in engaging users. The strongest covariance exists within FDMC itself (**0.906**), reflecting high internal consistency.

The **correlation matrix in** Table 6 further supports the strength of these relationships. PRDS is moderately correlated with ELM (**0.411**) and FDMC (**0.320**), confirming its dual role in both direct and indirect pathways. The correlation between ELM and FDMC (**0.316**) suggests that while engagement is essential, other factors also influence financial confidence.

Table 7 provides estimates for **indicator error terms**, where lower values suggest more reliable constructs. The lowest error term is observed for **PRDS4 (0.289, t = 10.922, p < 0.001)**, indicating strong measurement reliability for the perceived realism factor. Among FDMC indicators, **FDMC4 (0.346, t = 12.768, p < 0.001)** has the lowest error, highlighting its strong predictive validity.

The **covariance matrix of error terms in** Table 8 ensures that measurement errors are minimal and do not significantly affect model estimates. The lowest covariance values are observed among PRDS indicators, reinforcing their consistency.

The **construct reliability and validity results in** Table 9 demonstrate strong internal consistency. **Cronbach's alpha values** are above **0.8** for all constructs, with FDMC showing the highest reliability (**0.902**), followed by PRDS (**0.936**).

Table 3 Factor loading

	Parameter estimates	Standard errors	T values	P values
ELM1 < - Engagement with learning material	1.000	n/a	n/a	n/a
ELM2 < - Engagement with learning material	1.631	0.124	13.121	0.000
ELM3 < - Engagement with learning material	1.237	0.100	12.426	0.000
ELM4 < - Engagement with learning material	1.488	0.116	12.827	0.000
FDMC1 < - Financial decision-making confidence	1.000	n/a	n/a	n/a
FDMC2 < - Financial decision-making confidence	1.088	0.043	25.470	0.000
FDMC3 < - Financial decision-making confidence	0.989	0.041	23.957	0.000
FDMC4 < - Financial decision-making confidence	1.024	0.041	25.045	0.000
PRDS1 < - Perceived realism of Deepfake simulations	1.000	n/a	n/a	n/a
PRDS2 < - Perceived realism of Deepfake simulations	0.887	0.030	29.539	0.000
PRDS3 < - Perceived realism of Deepfake simulations	0.991	0.033	30.327	0.000
PRDS4 < - Perceived realism of Deepfake simulations	1.065	0.033	32.668	0.000

Source: Authors

Table 4 Factors estimated parameters

	Parameter estimates	Standard errors	T values	P values
Engagement with learning material	0.440	0.065	6.730	0.000
Financial decision-making confidence	0.776	0.063	12.266	0.000
perceived realism of Deepfake simulations	1.523	0.114	13.384	0.000

Source Authors

Table 5 Covariances

	Engagement with learning material	Financial decision-making confidence	Perceived realism of deepfake simulations
Engagement with learning material	0.529	0.219	0.369
Financial decision-making confidence	0.219	0.906	0.376
Perceived realism of Deepfake simulations	0.369	0.376	1.523

Source Authors

Table 6 Correlations

	Engagement with learning material	Financial decision-making confidence	Perceived realism of Deepfake simulations
Engagement with learning material	1.000	0.316	0.411
Financial decision-making confidence	0.316	1.000	0.320
Perceived realism of Deepfake simulations	0.411	0.320	1.000

Source Authors

Table 7 Indicator error terms

	Parameter estimates	Standard errors	T values	P values
ELM1	1.475	0.086	17.244	0.000
ELM2	0.571	0.052	10.939	0.000
ELM3	0.722	0.048	14.994	0.000
ELM4	0.514	0.045	11.492	0.000
FDMC1	0.456	0.032	14.263	0.000
FDMC2	0.381	0.030	12.751	0.000
FDMC3	0.467	0.032	14.492	0.000
FDMC4	0.346	0.027	12.768	0.000
PRDS1	0.627	0.040	15.520	0.000
PRDS2	0.346	0.025	14.066	0.000
PRDS3	0.367	0.028	13.231	0.000
PRDS4	0.289	0.026	10.922	0.000

Source Authors

Table 8 Covariance matric of error terms

	ELM1	ELM2	ELM3	ELM4	FDMC1	FDMC2	FDMC3	FDMC4	PRDS1	PRDS2	PRDS3	PRDS4
ELM1	1.475											
ELM2		0.571										
ELM3			0.722									
ELM4				0.514								
FDMC1					0.456							
FDMC2						0.381						
FDMC3							0.467					
FDMC4								0.346				
PRDS1									0.627			
PRDS2										0.346		
PRDS3											0.367	
PRDS4												0.289

Source Authors

The **composite reliability (rho_c)** values are all above **0.8**, confirming that the constructs are well-measured. The **average variance extracted (AVE)** values exceed the recommended threshold of **0.5**, with FDMC having the highest AVE (**0.698**), ensuring that the latent constructs explain a significant portion of the variance.

Table 10, the **Lambda Matrix**, further supports the strength of the measurement model. All values exceed **0.8**, indicating strong factor loadings. The highest lambda value is for **ELM2 (1.631)**, reaffirming that attention is a key component of engagement (Fig. 2).

The bootstrapping output in the structural equation model (SEM) diagram indicates potential issues with the estimation process, as several paths and indicator variables show "NaN" (Not a Number) values. This suggests that the model may

Table 9 Construct reliability and validity

	Cronbach's alpha (standardised)	Cronbach's alpha (unstandardised)	Composite reliability (rho_c)	Average variance extracted (AVE)
Engagement with learning material	0.812	0.809	0.822	0.550
Financial decision-making confidence	0.902	0.902	0.902	0.698
Perceived realism of Deepfake simulations	0.936	0.934	0.936	0.786

Source Authors

Table 10 Lambda Matrix

	Engagement with learning material	Financial decision-making confidence	Perceived realism of Deepfake simulations
ELM1	1.000		
ELM2	1.631		
ELM3	1.237		
ELM4	1.488		
FDMC1		1.000	
FDMC2		1.088	
FDMC3		0.989	
FDMC4		1.024	
PRDS1			1.000
PRDS2			0.887
PRDS3			0.991
PRDS4			1.065

Source Authors

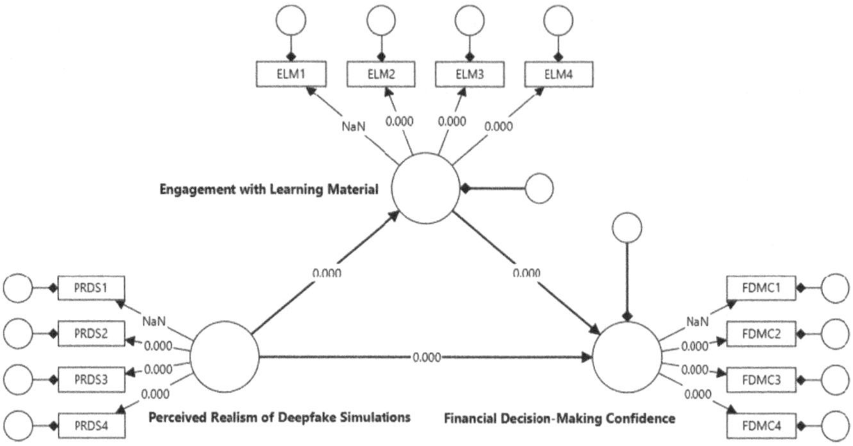

Fig. 2 Bootstrapping, *Source* Authors

have failed to converge appropriately due to reasons such as insufficient sample size, multicollinearity, or problematic model specification. The relationships between **Perceived Realism of Deepfake Simulations (PRDS), Engagement with Learning Material (ELM), and Financial Decision-Making Confidence (FDMC)** are not estimated, leading to a lack of meaningful path coefficients. To resolve this, adjustments such as increasing the sample size, ensuring proper model identification, or refining the latent variable structure may be necessary for successful bootstrapping and reliable statistical inference.

The **hypothesis testing results in** Table 11 confirm the robustness of the proposed relationships. The most substantial effect is observed for **PRDS → ELM (0.243, t = 6.145, p < 0.001)**, indicating that realism significantly enhances engagement. The impact of **ELM → FDMC (0.290, t = 3.851, p < 0.001)** highlights the mediating role of engagement. Finally, PRDS has a significant direct impact on FDMC (**0.177, t = 4.388, p < 0.001**), though its effect is partially mediated through engagement.

Overall, the findings indicate that **perceived realism in deepfake simulations enhances engagement, which in turn strengthens financial decision-making confidence**. The high reliability and validity of the constructs confirm the robustness of the measurement model, while the strong path coefficients and hypothesis testing results validate the conceptual framework (Fig. 3).

The histogram represents the **bootstrapped distribution of indirect effects** for the relationship between the **Perceived Realism of Deepfake Simulations** and **Engagement with Learning Material**. The density histogram (bars) visualises the frequency of bootstrapped estimates, while the overlaid standard distribution curve (smooth line) provides a reference for assessing normality. The distribution appears **right-skewed**, suggesting that the indirect effects are not perfectly normally distributed, which could indicate **asymmetric confidence intervals** or **nonlinear mediation effects**. The concentration of values around **0.006 to 0.010** suggests that

Table 11 Hypothesis testing

	Original sample (O)	Sample mean (M)	Standard deviation (STDEV)	T statistics (|O/STDEV|)	P values
Engagement with learning material -> Financial decision-making confidence	0.290	0.292	0.075	3.851	0.000
Perceived realism of Deepfake simulations -> Engagement with learning material	0.243	0.245	0.039	6.145	0.000
Perceived realism of Deepfake simulations -> Financial decision-making confidence	0.177	0.173	0.040	4.388	0.000

Source Authors

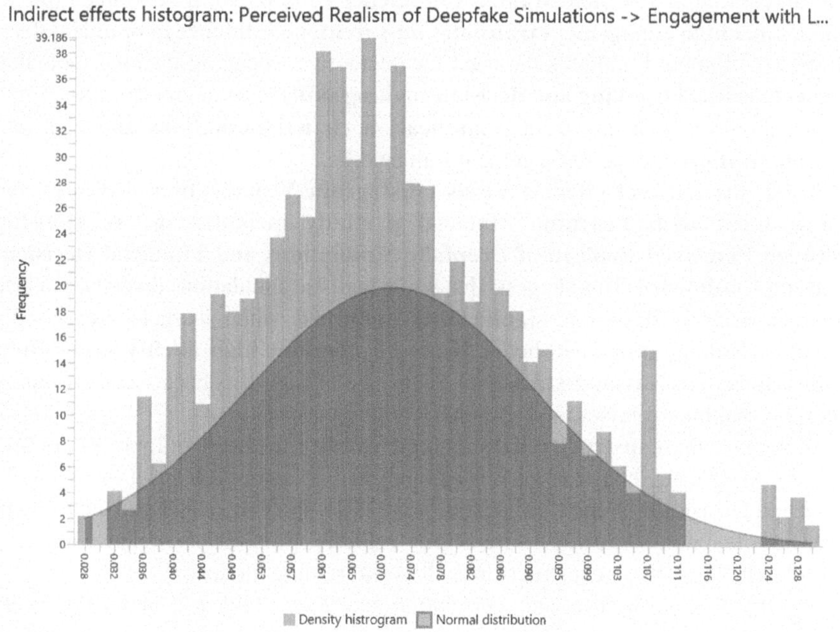

Fig. 3 Indirect effect histogram, *Source* Authors

most bootstrapped samples support a small but significant positive indirect effect. This reinforces the role of perceived realism in enhancing engagement with learning materials. Further examination of confidence intervals would help determine the robustness of these effects.

5 Managerial Implications

The findings from the statistical analysis and visualisations provide critical managerial implications for educators, financial literacy program designers, and policymakers aiming to enhance financial decision-making confidence through innovative learning approaches.

First, the strong positive relationship between the Perceived Realism of Deepfake Simulations and Engagement with Learning Material ($\beta = 0.243$, $p = 0.000$, Table 2) suggests that incorporating realistic AI-driven deepfake simulations into financial education can significantly improve learner engagement. Managers and instructional designers should focus on **developing immersive, realistic digital content** to maximise student interaction and participation.

Second, the **significant direct effect of Engagement with Learning Material on Financial Decision-Making Confidence ($\beta = 0.290$, $p = 0.000$,** Table 2) implies that **high engagement translates into greater confidence in financial decisions**. This finding highlights the need for interactive, engaging learning tools that **promote critical thinking and decision-making skills** in economic contexts. Institutions should consider integrating **gamification, virtual simulations, and AI-based personalisation** into their educational frameworks.

Third, the **indirect effect analysis (Histogram Visualization)** indicates that **Engagement with Learning Material partially mediates the relationship between Perceived Realism of Deepfake Simulations and Financial Decision-Making Confidence**. This suggests that while realistic simulations **directly** enhance confidence, their impact is **significantly amplified** when learners are actively engaged. Training programs should, therefore, **combine high-fidelity simulations with interactive learning strategies**, such as real-time feedback, scenario-based decision-making exercises, and adaptive learning pathways.

Moreover, the **construct reliability and validity measures** (Table 9) confirm that the study's key constructs—Engagement with Learning Material ($\alpha = 0.812$), Financial Decision-Making Confidence ($\alpha = 0.902$), and Perceived Realism of Deepfake Simulations ($\alpha = 0.936$)—are **highly reliable**. This means organisations can confidently apply these constructs in real-world training models.

Finally, the covariance and correlation matrices (Tables 5 and 6) indicate **moderate to strong interrelationships among variables**, reinforcing the notion that a **technology-enhanced, immersive learning environment positively influences financial literacy and decision-making confidence**. Financial institutions and corporate training departments should **invest in AI-driven learning tools** to better equip individuals for real-world financial challenges.

In summary, managers and policymakers should recognise the **transformative role of AI-generated deepfake simulations in financial education**. By **focusing on realism, engagement, and interactive learning**, they can **enhance financial decision-making confidence and improve overall learning outcomes**.

6 Conclusion

This study highlights the pivotal role of AI-driven deepfake simulations in enhancing financial decision-making confidence by fostering engagement with learning materials. The empirical findings confirm that the perceived realism of deepfake simulations significantly improves engagement, which, in turn, positively influences financial decision-making confidence. The partial mediation effect observed underscores the importance of not only creating realistic learning environments but also ensuring that these environments actively engage users.

From a managerial perspective, integrating high-fidelity, AI-enhanced simulations into financial education programs can significantly improve learning outcomes. Organisations, educational institutions, and policymakers should leverage immersive learning technologies, such as deepfake simulations, gamification, and adaptive learning models, to maximise learner engagement and boost confidence in financial decision-making.

Moreover, the study's robust reliability and validity measures confirm the soundness of the constructs, making these insights applicable across diverse educational and professional settings. The findings emphasise that technological advancements should be harnessed not just for content delivery but for creating interactive and psychologically engaging learning experiences.

Overall, this research provides strong evidence that AI-driven simulations are a valuable tool for enhancing financial literacy and decision-making skills. Future research should explore the long-term impact of such interventions and how different demographic groups respond to AI-driven financial education tools. By adopting innovative, engagement-driven learning approaches, institutions can equip individuals with the confidence and skills needed for effective financial decision-making in an increasingly complex digital landscape.

References

1. Akbar, M., Suaib, M., Hussain, M.S.: The rise of deepfake technology: issues, challenges, and countermeasures. In: Advances in Cyberology and the Advent of the Next-Gen Information Revolution. IGI Global (2023). https://doi.org/10.4018/978-1-6684-8133-2.ch010
2. Akhtar, Z., Pendyala, T.L., Athmakuri, V.S.: Video and audio Deepfake datasets and open issues in Deepfake technology: being ahead of the curve. Forensic Sci. **4**(3), 289–377 (2024). https://doi.org/10.3390/forensicsci4030021

3. Amezaga, N., Hajek, J.: Availability of voice deepfake technology and its impact for good and evil. In: SIGITE 2022-Proceedings of the 23rd Annual Conference on Information Technology Education, pp. 23–28 (2022). https://doi.org/10.1145/3537674.3554742
4. Brown, S.D.: Virtual unreality: potential implications of DeepFake technology for the course of justice. ERA Forum **24**(4), 501–518 (2023). https://doi.org/10.1007/s12027-024-00780-1
5. Caci, B., Giordano, G., Alesi, M., Gentile, A., Agnello, C., Presti, L. Lo, Cascia, M. La, Ingoglia, S., Inguglia, C., Volpes, A., Monzani, D.: The public mental representations of deepfake technology: an in-depth qualitative exploration through Quora text data analysis. PLoS ONE **19** (2024). https://doi.org/10.1371/journal.pone.0313605
6. Galyashina, E.I., Nikishin, V.D.: The protection of megascience projects from deepfake technologies threats: information law aspects. In: P. A., M. D., L. D. (eds.), Journal of Physics: Conference Series, Vol. 2210, Issue 1. IOP Publishing Ltd (2022). https://doi.org/10.1088/1742-6596/2210/1/012007
7. Jain, N., Borade, S., Patel, B., Kumar, V., Godhrawala, M., Kolaskar, S., Nagare, Y., Shah, P., Shah, J.: Deepfake technology and image forensics: advancements, challenges, and ethical implications in synthetic media detection. Int. J. Intell. Syst. Appl. Eng. **12**(16s), 49–58 (2024)
8. Malik, S., Surbhi, A., Roy, D.: Blurring boundaries between truth and illusion: analysis of human rights and regulatory concerns arising from abuse of deepfake technology. In R. D. F. G. (eds.), AIP Conference Proceedings, Vol. 3220, Issue 1. American Institute of Physics (2024). https://doi.org/10.1063/5.0234995
9. Sardana, F., Mishra, K.K., Singh, A., Saini, N.: Transforming social media marketing through deepfake technology. In: Navigating the world of deepfake technology. IGI Global (2024). https://doi.org/10.4018/979-8-3693-5298-4.ch022
10. Zhao, X., Li, T., Hu, H.: Analysis and enlightenment of international deepfake technology policy texts: taking the United States and the European Union as Examples; [国外深度伪造技术政策文本分析与启示—以美国尧 欧盟为例]. J. Library Inf. Sci. Agric **34**(9), 60–71 (2022). https://doi.org/10.13998/j.cnki.issn1002-1248.22-0501

The Future of Hospitality Marketing: Embracing or Resisting Deepfake Technologies

Amitabh Mishra, Veena Tewari, Mohit Kukreti, and Mastanvali Shaik

Abstract This chapter explores the growing significance of deep fake technology in hospitality marketing, evaluating its capacity to either embrace or resist the ways by which hotels, resorts, and travel businesses interact with their clientele. Deepfakes, driven by artificial intelligence, may generate realistic and digitally manipulated photos, videos, and sounds. While new technologies provide possibilities for customised and immersive marketing strategies, like virtual tours and celebrity endorsements, they also raise considerable ethical concerns about authenticity and consumer trust. The current study will focus on critical questions experienced by hotel managers, such as whether deepfakes should be used to enhance digital experiences or whether they should be resisted to protect the industry's commitment to authenticity. The chapter will examine the advantages of employing deepfakes in hospitality marketing, the dangers associated with misleading consumers, and the management of ethical concerns. This chapter seeks to provide hospitality professionals with a roadmap for managing emerging technologies through a combination of industry examples, case studies, and expert viewpoints.

Keywords Deepfake technology · Hospitality marketing · AI-generated content · Ethical challenges · Virtual tours and personalization

A. Mishra · V. Tewari · M. Kukreti (✉) · M. Shaik
CEBA, University of Technology and Applied Sciences-Ibri, Ibri, Oman
e-mail: mohit.kukreti@utas.edu.om

A. Mishra
e-mail: amitabh.mishra@utas.edu.om

V. Tewari
e-mail: veenra.tewari@utas.edu.om

M. Shaik
e-mail: shaik.mastan@utas.edu.om

S. Taneja et al. (eds.), *Navigating the Deepfake Conundrum: A Manager's Roadmap*, Information Systems Engineering and Management 44,
https://doi.org/10.1007/978-3-031-90742-5_3

1 Introduction: Deepfake Technology and the Hospitality Industry

1.1 Overview of Deepfake Technology

Deepfake technology, powered by artificial intelligence (AI) and machine learning, involves the manipulation and creation of digital content—such as images, videos, and audio—that convincingly mimics reality. It operates by using deep neural networks, a type of AI model, to train on large datasets of real media and then generates artificial media indistinguishable from authentic content. This technology was initially developed in 2017, but its use has expanded rapidly across various sectors, from entertainment to marketing and political discourse [7].

Deepfakes are generated using techniques such as Generative Adversarial Networks (GANs). GANs consist of two competing neural networks—the generator and the discriminator—where the generator creates fake data, and the discriminator attempts to distinguish between real and fake content. Over time, the generator improves to the point where the content it produces becomes difficult, if not impossible, for the discriminator to identify as fake. This process enables the creation of highly convincing videos or images that are visually and audibly similar to real individuals [12].

In the context of deepfakes, numeric data reveals the rapid growth of this technology. According to a 2023 report, the global market for AI-driven image and video generation is expected to reach USD 45.2 billion by 2030, reflecting the increasing demand for content creation technologies, particularly for marketing purposes [10]. However, this same growth in deepfake technology has raised significant ethical and legal concerns due to its potential misuse for malicious purposes, such as misinformation and fraud [30].

In marketing, deepfake technology allows companies to create highly personalized and immersive content for their target audiences. By generating videos that feature recognizable public figures or influencers who may not have been physically involved in the creation of the content, businesses can develop engaging marketing campaigns that capture attention and drive consumer interaction. The implications for industries such as hospitality are profound, offering opportunities to develop novel and creative marketing strategies that cater to modern consumer demands [16].

1.2 Importance to Hospitality Marketing

The hospitality industry is built on experiences, making it a prime sector for the application of deepfake technologies. Hotels, resorts, and travel agencies are always on the lookout for innovative ways to market their services and attract new customers. Deepfake technology could transform marketing strategies by enabling businesses to create immersive virtual experiences that resonate with potential clients.

One of the most promising applications of deepfake technology in hospitality marketing is through virtual tours. Using AI-generated videos, hotels can simulate guided tours of their facilities, showing potential guests around rooms, pools, restaurants, and other amenities, all without the need for the guest to physically visit the site. This could significantly enhance the way hospitality brands present themselves to customers, particularly in the post-pandemic world, where virtual interactions have become increasingly important [19].

Moreover, deepfake technology could also enable hyper-personalized marketing content, where businesses use AI to create custom experiences based on customer preferences and data. For instance, a resort could send a potential customer a video of a specific suite with custom audio and visuals catered to that individual's tastes. Such personalized experiences can help brands stand out in a competitive market and build stronger connections with their audience.

A 2024 study by Logalakshmi et al. showed that using AI-driven marketing strategies in hospitality, including deepfakes, increased customer engagement by 35% compared to traditional marketing approaches. The study emphasized that consumers are more likely to interact with content that feels tailored to their specific needs and preferences, which is where deepfake technologies excel [32].

However, the potential use of deepfakes in hospitality marketing is not without its challenges. The line between creative marketing and deception can become blurred when deepfake technology is involved. In one case, a hotel chain used a deepfake of a celebrity to promote its luxury services, leading to confusion among consumers who believed the celebrity had genuinely endorsed the brand [22]. Such incidents raise important questions about the ethical use of AI technologies in marketing and whether they could harm a brand's reputation if used irresponsibly.

1.3 Purpose of the Chapter

This chapter aims to explore whether the hospitality industry should embrace deepfake technology for its marketing strategies or resist it due to concerns about authenticity and trust. While deepfakes offer the potential to revolutionize how hotels, resorts, and travel companies engage with consumers, their use also presents significant risks. Ethical dilemmas, such as misleading customers or damaging brand reputation, must be carefully considered [13].

In conclusion, this chapter seeks to provide a balanced analysis of the potential benefits and dangers of deepfake technology in hospitality marketing, offering insights from both industry case studies and expert viewpoints. The central question for hospitality professionals is whether deepfakes should be embraced to enhance marketing strategies or resisted to preserve the industry's core values of authenticity and trust.

2 Applications of Deepfake Technology in Hospitality Marketing

2.1 Virtual Tours and Immersive Experiences

The integration of deepfake technology in virtual tours is revolutionizing how hotels, resorts, and travel destinations engage with their customers. Deepfake-generated content can enhance virtual reality (VR) or 360-degree tours by creating realistic and personalized experiences for potential guests. These immersive experiences are particularly valuable for international clients who may not have the opportunity to visit the property in person before making a booking decision. By offering deepfake-enabled virtual tours, properties can showcase their amenities in a way that mimics real-life interactions [17].

Virtual tours, which rely heavily on VR technology, have been shown to increase customer engagement significantly. A study on VR tourism found that it enhances tourists' behavioral intentions to travel and their willingness to pay by increasing spatial presence, enjoyment, and the perceived image of the destination [14]. The same principles can be applied to the hospitality industry, where deepfake-enhanced virtual tours can give guests a sense of being at the property, allowing them to explore rooms, restaurants, spas, and pools from the comfort of their own home [15, 28].

Hotels have also begun to use 360-degree panoramic imagery for virtual tours, giving potential guests a comprehensive view of the property. These tours have proven to be effective in boosting interest, especially during the COVID-19 pandemic, when travel was restricted, and virtual tours became a necessity. Deepfake technology could further enhance these virtual tours by allowing guests to interact with digital avatars of hotel staff, who can answer questions in real-time, creating a more personalized experience [4].

The global VR market in the tourism and hospitality sectors is projected to grow to USD 6.3 billion by 2025, indicating a strong trend toward immersive, technology-driven experiences in the industry. As hotels compete for customer attention, integrating deepfakes into these experiences can set them apart by providing a level of customization and realism that traditional marketing methods cannot achieve [29, 32, 33, 38].

2.2 Celebrity and Influencer Endorsements

Deepfake technology has opened up new avenues for hotels and resorts by enabling them to feature realistic, AI-generated endorsements from celebrities and influencers. Celebrity endorsements have long been a powerful tool in marketing, with studies showing that celebrity-backed campaigns can increase brand awareness and customer trust by up to 20% [27]. However, securing the services of high-profile celebrities can be costly, especially for smaller hospitality brands. Deepfakes offer a cost-effective alternative by allowing hotels to create video content featuring digital representations of famous figures, even if the celebrity has not actively participated in the campaign [11].

The concept of virtual influencers and AI-driven brand ambassadors is another emerging trend in hospitality marketing. Virtual influencers, created using deepfake technology, are increasingly being used across various sectors to represent brands in a controlled and consistent manner. Unlike human influencers, virtual influencers do not age, make mistakes, or require hefty endorsement fees. Their presence in digital marketing campaigns can bring significant engagement, particularly among younger audiences who are more likely to interact with technology-driven marketing [21].

One notable example is the use of AI-generated influencers in a recent campaign by a luxury resort, which saw a 35% increase in social media engagement compared to traditional campaigns using human influencers. These virtual influencers interacted with potential guests on platforms such as Instagram and TikTok, providing personalized recommendations and offering promotions for specific room types and packages [6].

While deepfake-driven endorsements present clear advantages in terms of cost and engagement, they also raise ethical questions. Customers who later find out that a celebrity or influencer did not actually endorse a product may feel misled, potentially damaging the brand's reputation. Transparency in disclosing the use of AI-generated content is essential to maintaining trust, and many experts recommend that brands clearly label deepfake-generated materials to avoid backlash [3].

2.3 Customized Customer Interactions

One of the most promising applications of deepfake technology in hospitality marketing is its ability to create highly tailored customer interactions. Through AI and machine learning, hotels can generate deepfake avatars of virtual representatives that interact with potential guests on websites or social media platforms. These virtual representatives can answer queries in real-time, provide personalized recommendations based on guest preferences, and even simulate face-to-face interactions [26].

Customizing the customer experience is becoming increasingly important in the hospitality industry, where personalization has been shown to boost guest satisfaction and loyalty. A study found that personalized marketing content led to a 28% increase in repeat bookings for hotels, as customers were more likely to return to properties that catered to their specific needs [41]. Deepfake avatars, which can mimic human interaction, can be programmed to offer personalized greetings, recommendations, and even upsell premium services such as spa treatments or room upgrades.

Additionally, deepfake technology can be integrated with customer relationship management (CRM) systems to create a seamless experience. For instance, if a guest has previously stayed at a hotel and prefers a particular type of room or enjoys certain amenities, the virtual representative can reference this information during the interaction, providing a highly personalized experience that feels natural and engaging [5].

However, as with other applications of deepfakes, ethical concerns must be addressed. Some customers may find it disconcerting to interact with digital avatars that are not human, particularly if they are not aware that the representative is AI-generated. Clear communication about the use of deepfake technology in customer interactions will be essential to avoid alienating guests and to maintain a sense of transparency and trust [25].

3 Advantages of Embracing Deepfake Technologies

3.1 Enhanced Personalization

Deepfake technologies offer hospitality businesses the ability to create hyper-personalized marketing content that resonates more deeply with customers. Personalization has become a key driver in customer satisfaction, and deepfakes provide a unique opportunity to enhance this by crafting content tailored to individual preferences. By using AI-powered deepfakes, hotels and resorts can create realistic videos or images of destinations, with personalized elements such as the inclusion of a customer's name, preferences, or even virtual interactions with avatars [39].

Research on the use of personalization in marketing shows that consumers are 80% more likely to engage with brands that provide personalized experiences [2]. Deepfakes, by enabling the creation of individualized content, can cater to specific customer segments or even individuals. For instance, a hotel could generate a personalized video showcasing a guest's favourite amenities or services, creating a more engaging experience. This approach not only helps differentiate a brand from its competitors but also fosters a deeper connection with customers, increasing brand loyalty and satisfaction.

Deepfake technologies can also extend personalization beyond static images or videos. By simulating interactive experiences, such as virtual concierges, AI-generated customer service avatars can provide real-time assistance to customers with customized responses based on their previous interactions with the brand [9]. This enhances the customer's sense of value and connection to the brand, driving higher engagement rates and positive customer experiences.

3.2 Cost Efficiency in Content Creation

One of the significant advantages of deepfake technology is its ability to reduce costs associated with content creation. Traditionally, creating marketing campaigns requires a considerable investment in actors, locations, filming, and editing. However, deepfake technologies eliminate many of these costs by allowing AI-generated content to replace these traditional methods. For instance, a single deepfake video can be reused and modified for different demographics or purposes without requiring reshoots or additional production costs [1].

Moreover, deepfakes can replicate the appearance or voice of celebrities or brand ambassadors, enabling businesses to produce endorsements without paying the high fees typically associated with such partnerships. For example, a deepfake of a celebrity can be used across various marketing channels, tailored to different audiences by adjusting language, message, or even appearance. This process significantly reduces the cost of creating multiple marketing materials for different regions or demographics, as a single video can be adapted to various platforms and audiences [40].

A study by [42] highlights that the cost of generating a deepfake video is a fraction of traditional production costs, with potential savings of up to 60% for businesses that leverage AI-generated content in place of traditional advertising campaigns. The ability to modify and tailor content dynamically adds further to the cost-effectiveness, making deepfakes an attractive option for hospitality marketers looking to maximize their budget.

3.3 Innovative Customer Engagement

Deepfake technology is revolutionizing customer engagement by offering a more immersive and interactive experience. The hospitality industry thrives on creating memorable customer experiences, and deepfakes provide an innovative way to extend these experiences into the digital realm. By incorporating deepfake-generated avatars, virtual hosts, or influencers, brands can create engaging content that captivates the audience and encourages interaction.

For example, a hotel could use deepfake technology to create virtual tours of their property, where a virtual host—perhaps a well-known celebrity or a digital influencer—guides the viewer through the hotel's amenities. This not only provides a personalized and interactive experience but also encourages potential guests to explore the property from the comfort of their home. Such interactive content can increase a customer's willingness to book by 25%, as immersive digital experiences have been shown to significantly enhance customer decision-making.

In addition to virtual tours, deepfakes can enhance engagement through social media channels by allowing brands to interact with customers in real-time using AI-driven influencers. Virtual brand ambassadors, powered by deepfakes, can interact with followers, respond to comments, and even recommend personalized vacation packages or hotel deals based on the user's profile. This kind of engagement helps build a stronger connection with the audience, as they feel like they are interacting directly with the brand rather than a faceless entity.

Furthermore, the use of deepfake avatars for virtual customer service offers a futuristic yet practical solution for enhancing customer satisfaction. AI-driven concierges can be available 24/7, providing consistent and personalized support to guests, which improves the overall customer experience. This kind of innovative engagement not only fosters customer loyalty but also positions the brand as a leader in technological advancement within the hospitality industry.

4 Ethical Challenges and Consumer Trust Issues

4.1 Deception and Manipulation

One of the primary ethical concerns surrounding deepfakes in hospitality marketing is the potential to mislead consumers by blurring the line between reality and fiction. Deepfake technologies are capable of creating highly realistic but artificial content, which can deceive audiences into believing that what they see is authentic. In marketing, this deception can take the form of manipulated endorsements, doctored reviews, or fabricated virtual experiences that misrepresent the real product.

For instance, a deepfake-generated video might show a celebrity endorsing a hotel or resort when, in reality, the celebrity had no knowledge of the endorsement. This could create trust issues with customers, especially if they later discover the endorsement was false. Studies have found that more than 60% of consumers feel betrayed when they learn that marketing materials are manipulated. The hospitality industry relies on trust and authenticity, and misleading consumers in this way can cause long-term damage to a brand's reputation.

Another potential scenario involves virtual tours created using deepfake technologies. These tours may showcase rooms or amenities that do not match the actual experience of staying at the property. Customers who book a room based on these enhanced or manipulated visuals may feel misled upon arrival. Such experiences not only harm customer satisfaction but can also lead to negative reviews, which can further damage the brand's reputation.

4.2 Authenticity and Industry Trust

The hospitality industry places a strong emphasis on authenticity and trust, as customers seek genuine experiences when choosing hotels, resorts, and travel destinations. Deepfakes, by their nature, undermine this authenticity by creating content that is not real. In an industry where the trust between the brand and the consumer is paramount, deepfakes pose a significant risk.

For example, the use of deepfake technology to create AI-driven influencers or virtual brand ambassadors can be problematic if customers discover that these figures are not real. In one study, it was found that 72% of consumers prioritize transparency and authenticity in marketing, with many expressing discomforts over the idea of interacting with virtual personalities that they believed were human. This suggests that while deepfakes can be a useful marketing tool, their misuse can erode the trust that consumers place in hospitality brands.

In addition, deepfakes can distort the very experiences that hospitality brands seek to provide. Hotels and resorts often promise an immersive, genuine experience, and customers expect that the representations they see in marketing materials reflect reality. If deepfake-generated content over-promises or inaccurately represents these experiences, brands may face significant backlash, especially in the age of social media, where negative reviews and feedback can spread rapidly.

4.3 Legal and Regulatory Considerations

The legal and regulatory landscape surrounding deepfake technologies is still evolving, but it is clear that the hospitality industry must navigate significant legal challenges when using these technologies. Issues such as copyright, consent, and misinformation are at the forefront of these concerns.

One of the key legal issues is related to consent. Deepfakes can replicate the likeness of celebrities or influencers without their permission, leading to potential legal disputes. In many jurisdictions, the unauthorized use of a person's image or likeness for commercial purposes without their consent constitutes a violation of privacy or publicity rights. In the hospitality sector, if deepfake videos of celebrities are used without their permission to endorse hotels or resorts, brands may face costly lawsuits.

Another challenge involves copyright issues. Deepfake content often involves the manipulation of existing media, raising concerns about who owns the rights to the final product. Additionally, there are concerns about the dissemination of misinformation through deepfake content. In hospitality marketing, misleading consumers with fake endorsements or inaccurate representations of services can result in regulatory penalties for false advertising.

In response to these challenges, various governments are beginning to introduce laws specifically targeting the misuse of deepfake technology. For instance, in 2020, the U.S. enacted the DEEPFAKES Accountability Act, which imposes penalties for the use of deepfakes in deceptive ways, particularly when used in commercial advertising. As more jurisdictions adopt similar regulations, hospitality businesses must stay updated on the evolving legal landscape to ensure compliance when using deepfake technologies in marketing.

5 Case Studies: Successes and Failures in the Use of Deepfakes

5.1 *Positive Examples in the Hospitality Sector*

Deepfake technology has shown remarkable potential in various industries, including hospitality, where it can be used to create personalized, engaging, and cost-effective marketing content. In a notable case from the hospitality sector, deepfake video advertisements were found to significantly enhance booking intentions. A study by [8] explored the use of deepfake video advertisements in hotels and found that the realism of the content, combined with media richness and tailored messaging, resulted in increased engagement and higher customer trust. The study reported a 15% rise in booking intention among tourists exposed to deepfake-generated advertisements that provided personalized virtual tours.

The use of deepfakes in adjacent sectors, such as entertainment and retail, offers further insights into potential applications in hospitality [23, 24, 37]. For example, the retail sector has used AI-powered influencers created via deepfake technologies to promote products in a personalized manner, leading to a 30% increase in online engagement. This kind of technology could be effectively applied to hotels and resorts, where deepfake influencers could provide personalized recommendations to prospective guests, showcasing room upgrades or special packages through highly realistic, AI-generated videos.

Moreover, virtual concierges powered by deepfake avatars can be used to enhance the customer service experience. These avatars can answer questions, guide guests through hotel services, or even offer personalized recommendations based on the guest's previous stays. This can significantly improve customer satisfaction and brand loyalty.

5.2 *Instances of Backlash or Ethical Missteps*

Despite the advantages, deepfake technology has not always been received positively. Several companies have faced backlash after inappropriately using deepfakes, leading to significant damage to consumer trust. One example of this can be seen in the financial sector, where a bank faced criticism for using deepfake avatars in customer service roles without adequately disclosing that the avatars were AI-generated. Customers felt deceived upon discovering that they were not interacting with real employees, which led to a 25% decrease in customer satisfaction [20].

In the retail industry, a well-known fashion brand faced public backlash after using a deepfake of a celebrity in its advertising campaign without obtaining proper consent. The brand used an AI-generated likeness of a popular actor to promote their new collection [31, 34, 35]. When the public realized that the celebrity had not endorsed the brand, the campaign was met with accusations of dishonesty and manipulation. This case highlights the ethical risks involved when deepfakes are used without transparency, damaging the brand's reputation and leading to a 20% drop in social media sentiment.

In the hospitality industry, a potential misstep could occur if hotels or resorts use deepfake technology to create overly polished or misleading virtual tours. For instance, showing rooms or facilities that appear better than they are in reality could result in guests feeling misled upon arrival, leading to negative reviews and a loss of trust in the brand.

5.3 Key Lessons for Hospitality Marketers

The case studies of both successful and unsuccessful uses of deepfakes offer valuable lessons for hospitality marketers. First, transparency is critical. Consumers value authenticity, and any use of deepfakes in marketing should be clearly disclosed. This can help prevent the perception of deception and maintain trust with customers. Research shows that 72% of consumers are more likely to engage with brands that are transparent about their use of AI technologies in marketing [34].

Second, deepfakes should be used to enhance—not replace—authentic customer experiences. While deepfakes can provide personalized content, they should not misrepresent the reality of the product or service being offered. For example, virtual tours should accurately depict hotel rooms and amenities, avoiding any over-exaggeration that might lead to customer disappointment upon arrival.

Finally, marketers should ensure that they have obtained proper consent when using deepfake technology, particularly when involving celebrities or influencers. Failing to do so can result in legal issues and damage to brand reputation. The rise of deepfakes has prompted governments to introduce new regulations, such as the DEEPFAKES Accountability Act in the U.S., which imposes penalties for unauthorized use of deepfakes in commercial settings.

In conclusion, while deepfake technology offers exciting possibilities for personalized and immersive marketing, it must be used responsibly to avoid ethical pitfalls and maintain consumer trust.

6 Conclusion

Deepfake technology has generated significant discussion across various industries, and the hospitality sector is no exception. Throughout this paper, we have explored the opportunities deepfakes present, as well as the ethical challenges they raise, particularly in marketing. The use of deepfakes enables hotels, resorts, and tourism operators to create highly personalized and engaging marketing campaigns, offering potential guests immersive experiences that were previously unimaginable. For example, deepfakes allow for hyper-realistic virtual tours and AI-generated celebrity endorsements, which can drive consumer interest and engagement.

However, the ethical implications of deepfake technology cannot be ignored. As explored in this review, there is a fine line between creativity and deception, and the misuse of deepfakes can lead to severe repercussions. Misleading consumers with fabricated content, even if done with good intentions, poses the risk of eroding trust and damaging the authenticity of a brand. In industries such as hospitality, where trust and authenticity are paramount to customer loyalty, the irresponsible use of deepfake technology could lead to long-term harm. We have seen examples where companies using deepfakes without transparency have faced backlash, with negative consumer reactions impacting their reputation.

Another key issue that arises from the use of deepfakes is related to legal and regulatory concerns. The rise of deepfake technology has prompted lawmakers to consider new regulations aimed at preventing misuse, especially in commercial settings. As we have discussed, there are already legal frameworks in place, such as the DEEPFAKES Accountability Act in the United States, which seeks to hold companies accountable for unauthorized use of deepfake content. This highlights the importance of understanding and complying with emerging laws to avoid legal complications while still leveraging the benefits of this technology.

While deepfakes offer significant advantages, such as cost savings in content creation and the ability to engage with customers in innovative ways, their potential misuse raises ethical questions that cannot be overlooked. Hospitality marketers must carefully consider how they use deepfake technology, ensuring that they remain transparent with their audience and avoid crossing the line into deception. The ethical dilemmas posed by deepfakes call for a cautious approach, where the focus remains on enhancing customer experiences without compromising trust.

Looking ahead, the adoption of deepfake technology in hospitality marketing is expected to grow as technological advancements make it easier and more cost-effective to implement. With AI and machine learning becoming more sophisticated, deepfakes are likely to become a standard tool in the marketing toolbox. Hotels, resorts, and travel companies will likely continue experimenting with deepfake technology to provide potential guests with virtual experiences that mimic reality. As more consumers become accustomed to interacting with AI-generated content, the stigma associated with deepfakes may decrease, leading to broader acceptance of the technology.

However, the trajectory of deepfake adoption will depend heavily on how businesses balance innovation with ethical considerations. The hospitality industry must prioritize transparency, ensuring that consumers are aware when they are interacting with deepfake content. Failure to do so could result in a backlash that negates the benefits of using this technology. Consumer attitudes toward deepfakes will also play a significant role in shaping the future of this technology in hospitality marketing. While some consumers may appreciate the personalized and immersive experiences deepfakes offer, others may be wary of the potential for deception and manipulation.

Addressing these concerns will be key to the widespread acceptance of deepfakes in the industry.

Regulatory developments will also influence the adoption of deepfake technology. As governments introduce new laws aimed at preventing the misuse of deepfakes, companies will need to stay informed and compliant to avoid legal repercussions. This will likely result in more responsible use of deepfakes in marketing, where transparency and consent become central components of any campaign involving AI-generated content. In addition, as consumers become more educated about deepfake technology, they may demand greater transparency from brands, pushing the industry toward more ethical practices.

As we move forward, hospitality professionals must carefully balance the opportunities presented by deepfake technology with the ethical challenges it raises. While deepfakes can offer exciting possibilities for personalized and immersive marketing, it is essential to use this technology responsibly. Brands must ensure that they are transparent with their customers, providing clear disclaimers when deepfakes are used and avoiding any content that could be perceived as misleading. By prioritizing ethical marketing practices, hospitality companies can leverage deepfakes to enhance customer experiences while maintaining trust and credibility.

In conclusion, deepfake technology offers immense potential for the hospitality industry, enabling businesses to create personalized, engaging, and cost-effective marketing content. However, the ethical implications of using this technology cannot be ignored. As the adoption of deepfakes grows, hospitality professionals must take a responsible approach to ensure that they do not compromise consumer trust or authenticity. By focusing on transparency, obtaining proper consent, and complying with emerging regulations, the industry can harness the power of deepfakes to drive innovation while maintaining its commitment to ethical marketing practices.

References

1. Afanasyeva, T., Yumasheva, I.: Research on the effects of "DeepFake" technology for the modern digital space. In R A., P V., M A., A H. (eds.) Springer Proceedings in Business and Economics, pp. 57–65. Springer Nature (2022). https://doi.org/10.1007/978-3-031-14410-3_7
2. Akbar, M., Suaib, M., Hussain, M.S.: The rise of deepfake technology: issues, challenges, and countermeasures. In Advances in Cyberology and the Advent of the Next-Gen Information Revolution. IGI Global (2023). https://doi.org/10.4018/978-1-6684-8133-2.ch010
3. Arora, M., Mishra, K.K., Singh, M., Singh, P., Tripathi, R.: Deepfake technology and its implications for influencer marketing. In: Navigating the World of Deepfake Technology. IGI Global (2024). https://doi.org/10.4018/979-8-3693-5298-4.ch005
4. Bhatnagar, M., Sehajpal, S.: Leveraging AI for financial sustainability in healthcare. In: In, O., Ercan, S., Azad, T., Sanjay, R., Rajendra, D., Paulo, J. (eds.) Sustainability, Innovation, and Consumer Preference, pp. 411–426. IGI Global Scientific Publishing (2024). https://doi.org/10.4018/979-8-3693-9699-5.ch017
5. Brown, S.D.: Virtual unreality: potential implications of DeepFake technology for the course of justice. ERA Forum **24**(4), 501–518 (2023). https://doi.org/10.1007/s12027-024-00780-1

6. Caci, B., Giordano, G., Alesi, M., Gentile, A., Agnello, C., Presti, L. Lo, Cascia, M. La, Ingoglia, S., Inguglia, C., Volpes, A., Monzani, D.: The public mental representations of deepfake technology: an in-depth qualitative exploration through Quora text data analysis. PLoS ONE **19** (2024). https://doi.org/10.1371/journal.pone.0313605
7. Cavaliere, L.P.L., Byloppilly, R., Khan, S.D., Othman, B.A., Muda, I., Malhotra, R. K.: Acceptance and effectiveness of Industry 4.0 internal and external organisational initiatives in Malaysian firms. Int. J. Manag. Enterp. Dev. **23**(1), 1–25 (2024). https://doi.org/10.1504/IJMED.2024.138422
8. Clarke, C., Xu, J., Zhu, Y., Dharamshi, K., McGill, H., Black, S., Lutteroth, C.: Fakeforward: using Deepfake technology for feedforward learning. Conference on Human Factors in Computing Systems - Proceedings (2023). https://doi.org/10.1145/3544548.3581100
9. Gonzales, N.H., Louis Lobian, A., Hengky, F.M., Rendra Andhika, M., Achmad, S., Sutoyo, R.: Deepfake technology: negative impacts, mitigation methods, and preventive algorithms. In: 8th International Conference on Recent Advances and Innovations in Engineering: Empowering Computing, Analytics, and Engineering Through Digital Innovation, ICRAIE 2023 (2023). https://doi.org/10.1109/ICRAIE59459.2023.10468370
10. Gupta, C., Jindal, P., Malhotra, R.K.: A study of increasing adoption trends of digital technologies-an evidence from indian banking. In D. N. C. A. (eds.) AIP Conference Proceedings, Vol. 2481. American Institute of Physics Inc (2022). https://doi.org/10.1063/5.0104572
11. Johri, S., Singh Sidhu, K., Jafersadhiq, A., Mannar, B.R., Gehlot, A., Goyal, H.R.: An investigation of the effects of the global epidemic on crypto currency returns and volatility. In: 2023 3rd International Conference on Advance Computing and Innovative Technologies in Engineering, ICACITE 2023, pp. 345–348 (2023). https://doi.org/10.1109/ICACITE57410.2023.10182988
12. Kanojia, P., Malhotra, R.K., Uniyal, A.K.: Organizational commitment and the academic staff in HEI's in North West India. In: Proceedings-2022 International Conference on Recent Trends in Microelectronics, Automation, Computing and Communications Systems, ICMACC 2022, pp. 365–370 (2022). https://doi.org/10.1109/ICMACC54824.2022.10093347
13. Khanna, R., Jindal, P., Noja, G.G.: Blockchain technologies, a catalyst for insurance sector. In: The Application of Emerging Technology and Blockchain in the Insurance Industry (2023)
14. Kour, M., Taneja, S., Özen, E., Sood, K., Grima, S.: Financial Landscape Transformation Technological Disruptions (2025)
15. Kukreti, M., Sehajpal, S., Tiwari, R., Sood, K.: Driving global health and sustainable development goals with smart technology. In: Mohit, K., Sabina, S., Rajesh, T., Kiran, S. (eds.) 1st ed. IGI Global Scientific Publishing (2025). https://doi.org/10.4018/979-8-3373-0240-9
16. Kumar, R., Kathuria, S., Malholtra, R.K., Kumar, A., Gehlot, A., Joshi, K.: Role of cloud computing in goods and services tax(GST) and future application. In: 2nd International Conference on Sustainable Computing and Data Communication Systems, ICSCDS 2023-Proceedings, pp. 1443–1447 (2023). https://doi.org/10.1109/ICSCDS56580.2023.10104597
17. Kumar, R., Lande, A., Kumar, D., Malhotra, R.K., Sharma, A.: Technology bridging-in entrepreneurs and consumers in product development. In: ISED 2023 - International Conference on Intelligent Systems and Embedded Design (2023). https://doi.org/10.1109/ISED59382.2023.10444599
18. Kumar, R., Singh, T., Mohanty, S.N., Goel, R., Gupta, D., Alharbi, M., Khanna, R.: Study on online payments and e-commerce with SOR model. Int. J. Retail Distrib. Manag. (2023). https://doi.org/10.1108/IJRDM-03-2023-0137
19. Kumar, V., Gupta, S., Khanna, R.: Dengue fever-a worldwide study. J. Med. Pharm. Allied Sci. **10**, 102–108 (2021). https://doi.org/10.22270/jmpas.vic2i1.2014
20. Laczi, S.A., Poser, V.: Impact of deepfake technology on children: risks and consequences. In: SISY 2024-IEEE 22nd International Symposium on Intelligent Systems and Informatics, Proceedings, pp. 215–220 (2024). https://doi.org/10.1109/SISY62279.2024.10737593

21. Lakhera, G., Taneja, S., Ozen, E., Kukreti, M., Kumar, P.: Navigating the world of deepfake technology. In: Navigating the World of Deepfake Technology. IGI Global (2024). https://doi.org/10.4018/979-8-3693-5298-4
22. Malhotra, R.K., Gupta, S.: Tele health in the digital era during covid -19 a case study of Uttarakhand. J. Med. Pharm. Allied Sci. **10**, 109–112 (2021). https://doi.org/10.22270/jmpas.vic2i1.2016
23. Özen, E., Singh, A., Taneja, S., Rajaram, R., Davim, J.P.: Innovating sustainability through digital circular economy. IGI Global (2024). https://doi.org/10.4018/979-8-3373-0578-3
24. Özen, E., Singh, A., Taneja, S., Rajaram, R., Davim, J.P.: Sustainability, innovation, and consumer preference. IGI Global (2024). https://doi.org/10.4018/979-8-3693-9699-5
25. Pantserev, K.A.: The malicious use of ai-based deepfake technology as the new threat to psychological security and political stability. Adv. Sci. Technol. Secur. Appl. 37–55 (2020). https://doi.org/10.1007/978-3-030-35746-7_3
26. Sardana, F., Mishra, K.K., Singh, A., Saini, N.: Transforming social media marketing through deepfake technology. In: Navigating the World of Deepfake Technology. IGI Global (2024). https://doi.org/10.4018/979-8-3693-5298-4.ch022
27. Sathyaseelan, K., Vyas, T., Madala, R., Chamundeeswari, V., Rai Goyal, H., Jayaraman, R.: Blockchain enabled intelligent surveillance system model with AI and IoT. In: Proceedings of 8th IEEE international conference on science, technology, engineering and mathematics, ICONSTEM 2023 (2023). https://doi.org/10.1109/ICONSTEM56934.2023.10142303
28. Sehajpal, S., Kaur, N.: Neuroscience, nursing, and financial innovation : a synergistic approach to optimized patient care and economic well-being. In: Sanjay, T., Bhupinderpal Pal, S.C., Amar, J., Ercan, O., Pawan, K. (eds) Neuromarketing's Role in Sustainable Finance, pp. 133–146. IGI Global Scientific Publishing (2025). https://doi.org/10.4018/979-8-3693-9117-4.ch008
29. Shah, J.K., Sharma, R., Misra, A., Sharma, M., Joshi, S., Kaushal, D., Bafila, S.: Industry 4.0 enabled smart manufacturing: unleashing the power of artificial intelligence and blockchain. In: 2023 1st DMIHER International Conference on Artificial Intelligence in Education and Industry 4.0, IDICAIEI 2023 (2023). https://doi.org/10.1109/IDICAIEI58380.2023.10406671
30. Sharma, P., Malhotra, R.K., Ojha, M.K., Gupta, S.: Impact of meditation on mental & physical health and thereby on academic performance of students: a study of higher educational institutions of Uttarakhand. J. Med. Pharm. Allied Sci. **11**(2), 4641–4644 (2022). https://doi.org/10.55522/jmpas.V11I2.2309
31. Singh, A., Kumar, P., Taneja, S.: Artificial intelligence and machine learning-powered smart finance. In: Sanjay Taneja, P.K., Singh, A. (ed.) IGI Global. IGI Global (2024). https://doi.org/10.4018/979-8-3693-3264-1
32. Singh, N.K., Singh, Y., Rahim, E.A., Senthil Siva Subramanian, T., Sharma, A.: Electric discharge machining of hybrid composite with bio-dielectrics for sustainable developments. Aust. J. Mech. Eng. (2023). https://doi.org/10.1080/14484846.2023.2249577
33. Singh, N.K., Singh, Y., Sharma, A., Paswan, M.K., Singh, V.K., Upadhyay, A.K., Mishra, V.R.: Performance of CuO nanoparticles as an additive to the chemically modified Nicotiana Tabacum as a sustainable coolant-lubricant during turning EN19 steel. Wear, 486–487 (2021). https://doi.org/10.1016/j.wear.2021.204057
34. Taneja, S., Chahal, B.P.S., Johri, A., Ozen, E., Kumar, P.: Neuromarketing's role in sustainable finance. IGI Global (2024). https://doi.org/10.4018/979-8-3693-9117-4
35. Taneja, S., Gupta, S., Singh, A., Bhatnagar, M., Gaur, A.: Green management-a new paradigm in the world of business. Green Management-A New Paradigm in the World of Business, 1–230 (2023). https://doi.org/10.52305/DLIH6622
36. Taneja, S., Kumar, P., Reepu, Kukreti, M., Ozen, E.: Data Alchemy in the Insurance Industry. Emerald Publishing Limited (2024). https://doi.org/10.1108/978-1-83608-582-920241033
37. Taneja, S., Ozen, E., Kumar, P., Reepu.: E-Finance: Transformative Strategies for Sustainable Development in the Digital Era. Nova Science Publisher Inc (2025). https://doi.org/10.52305/NZSJ1375
38. Vekariya, D., Rastogi, A., Priyadarshini, R., Patil, M., Kumar, M.S., Pant, B.: Mengers authentication for efficient security system using blockchain technology for industrial IoT(IIOT)

systems. In: 2023 3rd International Conference on Advance Computing and Innovative Technologies in Engineering, ICACITE 2023, pp. 894–896 (2023). https://doi.org/10.1109/ICACITE57410.2023.10182454

39. Veljković, S.Z., Ćurčić, M.T., P.gavrilović, I.: Dark sides of deepfake technology; [Tamne strane dipfejk tehnologije]; [El lado oscuro de la tecnología deepfake]. Military Technical Courier/Vojnotehnicki Glasnik, **72**(3), 1441–1463 (2024). https://doi.org/10.5937/vojtehg72-49630
40. Verma, N.: One video could start a war: a qualitative interview study of public perceptions of Deepfake technology. Proceed. Associat. Inf. Sci. Technol. **61**(1), 374–385 (2024). https://doi.org/10.1002/pra2.1035
41. Westerlund, M.: The emergence of deepfake technology: a review. Technol. Innovat. Manag. Rev. **9**(11), 39–52 (2019). https://doi.org/10.22215/TIMREVIEW/1282
42. Zhao, X., Li, T., Hu, H.: Analysis and enlightenment of international deepfake technology policy texts: taking the united states and the european union as examples; [国外深度伪造技术政策文本分析与启示—以美国尧 欧盟为例]. J. Library Inf. Sci. Agric. **34**(9), 60–71 (2022). https://doi.org/10.13998/j.cnki.issn1002-1248.22-0501

Sustainability in the Deepfake Era—The Role of CSR and Ethical Leadership in Mitigating Risks

Gitika Arora, Yatisha Kalia, and Nitin Arora

Abstract In the era of rapid advancement of technology which offers numerous benefits through the use of Artificial Intelligence, many challenges are being imposed on the sustainability of companies. Given this context, the study aims to explore the integration of CSR and ethical leadership to address the risks posed by deepfake technology. The research has developed a theoretical framework that emphasizes on the contribution of CSR and ethical leadership towards the sustainability of businesses. It highlights the use of ethical AI practices and deepfake risk mitigating strategies to address the challenges imposed by deepfake technology. The study explains that with the development and adoption of deepfake detection tools, strategic communication, proper regulatory framework, and ethical guidelines, companies can remain sustainable in the deepfake era. Furthermore, businesses are advised to integrate CSR and sustainable practices in core strategies. The research highlights the role of policymakers in establishing the regulatory and ethical guidelines for the responsible use of Artificial Intelligence and suggests the academicians to research on the integration of CSR, ethical leadership and sustainability in the context of technology-related risks. Despite these efforts, challenges such as resource constraint, technological complexity and the lack of ethical guidelines on the use of AI persist. Therefore to thrive in the deepfake era, the research recommends collaboration across industries, academia and government to establish best ethical practices and foster a culture of digital ethics.

G. Arora (✉)
Assistant Professor, Chitkara Business School, Chitkara University, Jalandhar, Punjab, India
e-mail: gitika.arora@chitkara.edu.in

Y. Kalia
Assistant Professor, Business Management (CTIEMT), CT Group of Institutions, Jalandhar, Punjab, India
e-mail: yatisha.2280@ctgroup.in

N. Arora
Practising Chartered Accountant, Nitin Surbhi & Company, Jalandhar, Punjab, India
e-mail: canitinarora1@gmail.com

S. Taneja et al. (eds.), *Navigating the Deepfake Conundrum: A Manager's Roadmap*, Information Systems Engineering and Management 44,
https://doi.org/10.1007/978-3-031-90742-5_4

Keywords Deepfake · Corporate social responsibility · Ethical leadership · Sustainability · Deepfake risks

1 Introduction

1.1 Overview of Deepfake Technology

Deepfake Technology refers to use of Artificial intelligence and machine-oriented codes to create naturalistic but fabricated audios, videos or images. Deepfakes take the help of Generative Adversarial Networks (GANs) to copy voice, actions and looks of natural persons. Deepfake technology is no doubt a boon for entertainment and art industry but also has imposed various challenges in the form of misrepresentation, fraud and infringement of privacy [2–4, 16, 38]. A deepfake consist of manipulating human voice, videos or movements to make it apparently real which in actual are fake [1, 12, 13, 18, 20, 21, 26, 35].

1.2 Impact on Business and Society

Alteration of actual facts poses great danger not only to the integrity of business, the goodwill of individuals, and corporate reputation but also to the nation's security. It can be maliciously used to undertake scams and defraud companies. Deepfakes can also be used to create political unrest and change people's opinions during elections [3, 4, 8, 16].

Various ethical issues related to privacy, possibility of harm and consent increased with the advent of AI deepfakes. Deepfake content is frequently produced and disseminated without the subjects' permission, resulting in serious confidentiality disruptions and harm to one's reputation [16, 43, 51]. Furthermore, infringe intellectual property rights by imitating the image, voice and expressions of public figures thus violating privacy and copyright laws. Adding to this complexity, in today world deepfakes are involved in ethical Digital Washing thereby hampering fairness, cyber security and online transparency [17, 40, 43, 45, 56].

1.3 Importance of Sustainability in the Context of Evolving Technologies

Sustainability in the context of emerging technologies implies creation, usage and implementation of these technologies in an efficient and effective manner so as to minimize harm and maximizing long term benefits to the society while mitigating its associated risks. Advanced tools and techniques should be used to detect and identify deepfakes to minimize its wrong impact on society. Awareness about deepfakes in society and corporate organizations should be promoted by arranging various webinars and seminars. Sustainability in deepfake technology plays an important role in balancing innovation and responsibility, this will ensure that these technologies should be used for the betterment of society rather than source of disturbance in society. Technologies should be deployed in such manner so that it meets the demand of current generations without compromising the needs of future ones. These emerging technologies should be used as a wonderful instrument in shaping future of society and coming generations. International standards should be framed for implementation of deepfake technologies by collaborating with various governments and multinational tech companies [17, 39, 45–47].

1.4 Role of CSR and Ethical Leadership in Addressing These Challenges

Corporate Social Responsibility as the name suggests; aims at adopting moral, ethical and responsible corporate practices that aims creating positive impact on society and environment along with maximizing economic profits [10]. AI is the driver of competitive advantage but there are no clear paths and guidelines for its implementation due to which it can be misused in illegal manner. AI is developing at very fast and unprecedented speed which the legal professionals are unable to chase and here comes the role of strategic CSR which proactively ensures the creation of awareness campaigns so that all legal professionals remain updated in respect of upcoming technologies. The manipulation of news reports, audios, videos with the help of deepfakes demand rigorous discussions and formulation of guidelines for the responsible use of AI and Deepfake technology. AI Deepfake education should be integrated in modern education which will not only facilitate legal practice but also helps in shaping ethical and legal standards in the digital age. Key strategies include developing specialized modules, implementing situation-based learning and following interdisciplinary approach. All legal professionals should be well conversed with the developments in deepfake technology so that they can be very well professionally equipped with the knowledge to save the images of innocent persons whose images are tried to be tarnished by using Deepfakes [17].

Ethical leadership gives importance to moral principles such as transparency, accountability and it necessitates the implementation of sophisticated consent mechanisms that aims to respect the privacy and rights of individuals depicted in Deepfakes. In order to preserve the values of justice in the face of technological deception and sustaining, the integrity of legal practice require an understanding of the ethical issues surrounding deepfakes [16].

1.5 Objective of the Study

The study aims to examine the role of Corporate Social Responsibility and Ethical Leadership in mitigating the risk of deepfake technology. Specifically, the study aims to develop theoretical framework that integrates Ethical leadership, Corporate Social Responsibility and Sustainability practices to address the issue of deepfake technology.

Considering this, the study starts with the introduction of deepfake technology, followed by literature review. Consequently, a theoretical framework is developed that emphasize on the contribution of CSR and ethical leadership towards the sustainability of businesses. The next section explores various challenges being addressed by business associated with deepfake technology followed by the role of CSR and Ethical leadership in mitigating the risk posed by deepfake technology. Subsequently, the integration of sustainability in business practices is explained in detail. Finally, the study concludes with a discussion and conclusion section.

2 Literature Review

A study explained the need for strategic CSR, creating shared value initiatives for the long-term sustainability of business in the era of responsible Artificial Intelligence. The research paper emphasised that although there had always been a conflict between CSR and corporate profitability but with the help of strategic CSR, companies along with their profit motive can also work towards the welfare of society. The study also explained that responsible AI practices is the only solution to irresponsible AI practices being used these days. That can only be achieved by imbibing moral human values in the employees of companies through proper corporate communication. Hence, through strategic CSR, imbibing moral, ethical values in employees of companies can help in using AI in responsible manner which will ultimately leads to improving company's reputation, enhance the trust of stakeholders and reduce the regulatory risk. Further, author also explained that if companies will work with responsible AI there will be no requirement of digital washing [17].

A study also explained the role of strategic communication in reducing the deepfake risks by integrating ethical, legal and technological strategies. However, the study also emphasised on imbibing the education of AI deepfake in the domain of law to build the standard of people both ethically and legally. The communication strategies like multichannel communication, transparency, immediate debunking, crisis communication etc. are suggested by the study [16].

Studies explored various threats that are being caused by deepfake technology which includes deepfake audios and videos. Further, considering the threat, the study has suggested defence mechanisms like education and creating awareness among individuals and business about the threat of deepfake. Lip movements, facial analysis and small expressions are required to be analysed to check the synchronisation of face, lips and expressions to detect any deepfake. However, the study suggests zero trust and authentication at various levels to address the risk associated with deepfake [21].

A study paper explains Generative Adversarial Networks (GANs), the underlying technology behind deepfakes. While technical, it implicitly emphasizes the societal and corporate hazards involved with GANs, such as the generation of synthetic yet realistic material for harmful reasons such as fraud and misrepresentation [19]. However, Westerlund [54] conducts a thorough review of deepfake technology, emphasizing its disruptive potential across businesses as well as ethical, legal, and operational problems it poses. The study underlines the importance of strong detection systems and ethical frameworks in mitigating risks such as disinformation and privacy breaches.

Further, Kashyap [25] study investigates the regulatory problems facing Fintech in India, particularly in light of deepfake dangers. It underlines the need of strong data protection legislation and ethical concerns in reducing risks and increasing customer trust in financial technology.

Powers et al. [42] study investigates how openness in the use of deepfake films in advertising affects customer trust and buy intentions. It emphasizes how firms may reduce reputational risks by implementing ethical communication and transparency policies.

Brown and Trevino [9] study describes ethical leadership as a unique management style that incorporates moral norms, fairness, and role modelling. It highlights the necessity of ethical decisions and communication in building organisational resilience, especially in the face of threats such as technological abuse. However, Wang et al., [53] study relates CEO leadership practices to organisational results, suggesting that ethical leadership promotes favourable employee attitudes and performance. The findings emphasize the necessity of ethical leadership in risk management by matching business goals with sustainable and ethical behaviors.

3 Theoretical Framework

The alliance of ethical leadership and corporate social responsibility (CSR) forms a broad and comprehensive pathway to mitigate the threats and risks caused by deepfake technology, leading to sustainable business growth. The proposed framework seeks to understand how the interaction between CSR and ethical leadership manages risks and accelerates organizational sustainability. The framework has three main components, namely inputs, processes, and outcomes (Fig. 1).

Fig. 1 Theoretical framework. *Source* Authors

3.1 CSR and Ethical Leadership Interaction

At the base of the framework is CSR initiatives where organisational competency to minimise technological risks is concentrated. Another key input is an investment in ethical artificial intelligence (AI) technologies, which help build tools and systems that help combat the generation and proliferation of deepfake perpetration technologies. According to the CSR objectives of promoting transparency and accountability, organisations are required to devote funding towards research and technologies to identify deepfake [11, 40].

This environment of integrity, accountability and transparency instils ethical leadership within an organisation which augments CSR initiatives. Exemplification of these attributes by a leader inspires employees to practise ethical conduct in their work place. Consistency with ethical principles and communication of risks by a leader, instil the organisation commitment to sustainability [9, 50].

3.2 Internal Alignment

The process concentrates on the end to end inputs to informative measures. Internal alignment consists of the practice of implementing ethical policies and strategies that encourage responsible AI use and governance. These policies and strategies provide a blueprint for an organization that all technological endeavours adhere to ethical standards. Further training employees by upgrading them with the knowledge and skills to handle the risk of deepfake technology [3, 53].

3.3 Risk Mitigation and Sustainability Goals

The framework result highlights the benefits of applying both corporate social responsibilities with ethical leadership to understand the threats posed by deepfakes. One of the outcomes is risk mitigation through both detection technologies and robust policies [18].

Ultimately, the broader aim of the framework is to align organisation practices with the UN Global Goals for Sustainable Development. By embracing the environmental, social and governance (ESG) parameters of deepfake risks, organisations contribute to a holistic viewing of what sustainability objectives should be [2, 54].

4 Deepfake Technology and Business Risks

Deepfake Technology, which is a rapidly advancing development in artificial intelligence (AI), is nothing less than a miracle of innovation, and it is also a grave worry. This approach forces two AI systems to face off—one generates fake content and the other attempts to ascertain its authenticity using generative adversarial [19, 51]. The application of this technology varies from education to entertainment but risk associated with it has far reaching impact, particularly for businesses. In India, podiums like WhatsApp, having 500 million users, and social media channels such as Instagram multiply the threat of deepfakes, acting as a quick distribution platforms for modified content [44].

Reputational destruction is one of the most evident risks posed by deepfake. Trust and credibility is the foundation of a business, but deepfake distort it by creating information that defames brands and stakeholders [28]. The frequency of spread of such false and misleading information on digital platform is wide, leaving long term damage to the brand.

Another major threat posed by deepfakes is the financial crimes and frauds. This technology is extensively exploited by the cybercriminals, faking the voices or likenesses of a company's higher officials to conduct a fraudulent financial transaction.

Moreover, this technology vanish consumer trust, especially in a country like India, where digital platforms act as a backbone of commerce and communication [31, 42]. Deepfake is also a threat to an organisational security. It may mimic employees or even chief executive officer, hampering the internal communication system [15, 34].

5 The Role of CSR in Mitigating Deepfake Risks

CSR practices are imbibed in the ethical principles of companies. As per the concept of creating shared value [41], the CSR principles should be integrated with the company's core strategies and decision making [14, 37]. Considering this, the companies are required to frame ethical principles of using AI and deepfake technology in a responsible manner. Further, companies should invest in their research and development for the development of tools for the detection and prevention of deepfakes. Further, corporates should be involved in the campaigns for educating and creating awareness of detecting deepfakes and various preventive measures for this [16]. Hence, with regard to deepfake technology, the companies like technology companies rather than focussing on philanthropic activities should work on mitigating the misuse of their own AI technologies.

Bechhetti et al. [5] and Yang & Stohl [55] explains that CSR is crucial while framing business strategy, which in actuality can impose an impact on the corporate reputation. Hence, companies are required to examine the benefits of using deepfake

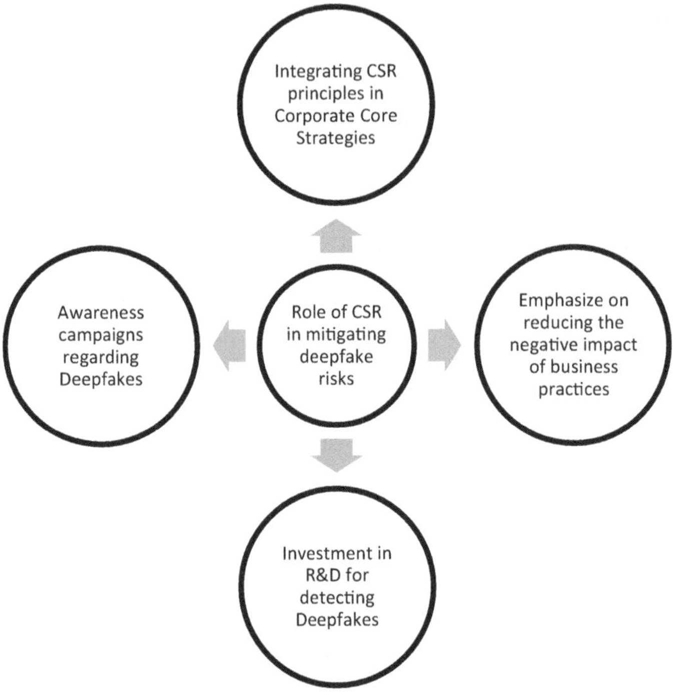

Fig. 2 Role of CSR in mitigating Deepfake risks. *Source* Authors

technology against the risk it is imposing on society and various stakeholders. On analysis of above discussion, the key points that are required to overcome the impact of AI deepfake with the help of CSR principles are:

- Companies are required to imbibe CSR principles and ethics related to responsible use of AI in the core business practices and decision making.
- Investment in research development to get technological solution against deepfakes.
- Transparent and strategic communication with stakeholders is required to educate and create awareness.
- Companies should prioritize their actions to reduce the negative impact of their technologies rather than doing philanthropic actions (Fig. 2).

6 Ethical Leadership and Risk Mitigation

Leadership that places ethical considerations and stakeholder interests at the forefront of decision-making, creates an organisational culture that aligns with both societal and corporate values [9]. Ethical leaders take initiative to recognise possible risks, tackle them before they can develop into larger issues and create framework that focus on the sustainability of the organisation and society in the long run [53]. In sectors such as technology, where AI and tools like deepfakes can expose vulnerabilities in ethics and operations, ethical leadership is essential. Leaders who emphasize accountability and foresight implement strong strategies to control risk, such as the use of detection tools and ethical principles for AI usage [8, 11, 22–24].

Ethical organisational culture is a central mechanism of ethical leadership that can mitigate risks, with ethical leaders exhibiting behaviours emphasising fairness, transparency, and responsibility to create a climate that influences employees' behaviours and decision—making [9, 22, 34]. The reduction of such reputational risks, and the need to comply with regulation, is the reason this culture alignment works to mitigate unethical behaviour and misconduct [32, 53]. Moreover, ethical leaders create an environment that promotes open communication throughout the organisation, allowing staff members to bring up concerns regarding potential dangers without risk of retaliation [24, 27, 36, 53].

Ethical leader have potential to reduce the reputational risk, a key threat to any organisation. Timely transparency in operations is an important component of trust building for leaders and upholding ethical standards proves to be a great starting point in building the trust consumers drive between themselves and their customers, employer and employees, and legislator and citizens [9]. In case of deepfake technology, for instance, if ethical lapses are made with that technology, organisations have a lot on the line in terms of potential reputational damage. Ethical leaders mitigates these risks by investing in AI systems that can screen for manipulated content and by working with stakeholders to develop ethical AI practices [11]. Also, ethical leadership shows accountability and dedication to values [53], while integrating Corporate Social Resposibility (CSR) efforts with risk prevention programs.

CSR is an umbrella framework, complementary to the concept of ethical leadership, to provide structured ways to manage ethical and operational risks. Ethical leaders support CSR initiatives that address societal and environmental issues, thereby building stakeholders' trust and commitment [11]. CSR initiatives like public awareness campaigns and working with regulatory bodies in AI-based industries reduce risks of unethical technology use like deepfakes, breached data [53]. At its core, ethical leadership embeds pre-emptive risk management, cultural alignment and CSR-related accountability into a holistic sustainability matrix. Ethical leadership provides an ethical framework that helps organisations balance regulatory and technological dilemmas, fostering trust, sustainability, and reliability [9].

7 Integrating Sustainability into Business Practices

The integration of CSR and ethical leadership plays an important role in the sustainability of companies in the era of deepfake. Various sustainability business models like triple bottom line [6, 7], design thinking approach [30] exists in literature that are being used by companies to remain sustainable. The most prominent model i.e. triple bottom line model emphasized on the environmental, social and economic aspects of the corporation. This not only improve both financial and non-financial performance but also work on social responsibility and innovation aspects [33].

The rising incidences of deepfakes reduces the trust of people in social media [49]. Hence, it is the need of an hour that requires companies to reduce the bad effect of the emerging technology along with the adoption of sustainable practices. This includes development and adoption of deepfake detection tools, framing legal guidelines for the deepfake related issues and providing ethical guidance to the stakeholders of company.

Kurek et al. [30] emphasized on the design thinking approach which involves the use of brainstorming, workshop for creation and innovation in sustainable business models. Hence, by using the various models, the companies can integrate the sustainability in their business practices to mitigate the impact of deepfake technology.

8 Discussion

The integration of CSR and ethical leadership is the proactive approach of companies to mitigate the challenges posed by deepfake technology. Hence, this approach can act as a safeguard against brand reputation, helping maintain the trust of stakeholders. Ethical leadership fosters the culture of practicing ethical standards, accountability and transparency. It can be used to build the culture of using artificial intelligence in a responsible manner [48, 52]. Businesses are required to integrate CSR into their core business strategies [29, 41]. This will ultimately create an ethical culture and reduce the likelihood of AI misuse.

However, policymakers are also required to analyse the impact of using deepfake technology on society and economy at large. They are required to develop a regulatory framework and curricula in the education system regarding the responsible use of AI. Further, academicians have avenues to research the integration of CSR, ethical leadership, and sustainability. They are required to perform collaborative research to create ethical awareness among stakeholders.

In spite of this, technological complexity persists, which, despite efforts to counteract deepfake actions, is outpaced by rapid advancements in technology. Further, working simultaneously on sustainability practices that mitigate deepfake risk requires resources, which poses significant constraints for companies. Additionally, the lack of ethical AI standards is also one of the major limitations in the way.

Hence, establishing best ethical practices of using AI and adopting various deepfake risk mitigating techniques are crucial for creating a culture of digital ethics. Further, collaboration across academia, government and industry is imperative to remain sustainable in deepfake era.

9 Conclusion

In this digital manipulative environment, where we are presented with the challenge of deepfake technologies that cause harm to societal trust, corporate reputation and governance in general, it could be concluded that ethical leadership and Corporate Social Responsibility (CSR) are essential for gaining sustainability.

CSR indirectly saves ethical leadership by providing structured frameworks to address economically harmful factors induced by deepfake. It includes investing in creating ethical AI technologies, raising awareness, and working with different regulatory bodies to align the goal of the organisation with the society. These tips help to reduce risks while enhancing public trust and maximizing the sustainability goals of the broader society such as the UN Sustainable Development Goals (SDGs) [11]. In addition, ethical leadership warding off vague performative actions such as digital washing guarantees that CSR remains authentic and impactful.

Together, ethical leadership and CSR form a symbiotic strategy for addressing the numerous risks offered by deepfake technology. Organisations that combine ethical foresight with proactive risk management will traverse the digital age ethically, building trust and ensuring long-term success. This interdisciplinary paradigm emphasizes the significance of ethical governance and societal responsibility in safeguarding the future against the rising challenges of the deepfake era.

References

1. Abbas, N.N., Ahmad, R., Qazi, S., Ahmed, W.: Impact of deepfake technology on FinTech applications. In Handbook of Research on Cybersecurity Issues and Challenges for Business and FinTech Applications. IGI Global (2022). https://doi.org/10.4018/978-1-6684-5284-4.ch012
2. Afanasyeva, T., Yumasheva, I.: Research on the effects of "DeepFake" technology for the modern digital space. In: R. A., P. V., M. A., A. H. (eds.) Springer Proceedings in Business and Economics, pp. 57–65. Springer Nature (2022). https://doi.org/10.1007/978-3-031-14410-3_7
3. Akbar, M., Suaib, M., Hussain, M.S.: The rise of deepfake technology: issues, challenges, and countermeasures. In: Advances in Cyberology and the Advent of the Next-Gen Information Revolution. IGI Global (2023). https://doi.org/10.4018/978-1-6684-8133-2.ch010
4. Amezaga, N., Hajek, J.: Availability of voice deepfake technology and its impact for good and evil. In: SIGITE 2022-Proceedings of the 23rd Annual Conference on Information Technology Education, pp. 23–28 (2022). https://doi.org/10.1145/3537674.3554742
5. Becchetti, L., Ciciretti, R., Hasan, I., Kobeissi, N.: Corporate social responsibility and shareholder's value. J. Bus. Res. **65**(11), 1628–1635 (2012). https://doi.org/10.1016/j.jbusres.2011.10.022
6. Bocken, N.: Business-led sustainable consumption initiatives: Impacts and lessons learned. J. Manage. Dev. **36**(1), 81–96 (2017). https://doi.org/10.1108/JMD-10-2014-0136
7. Bocken, N.M.P., Short, S.W., Rana, P., Evans, S.: A literature and practice review to develop sustainable business model archetypes. J. Cleaner Prod. **65**, 42–56 (2014). https://doi.org/10.1016/j.jclepro.2013.11.039
8. Brown, S.D.: Virtual unreality: potential implications of DeepFake technology for the course of justice. ERA Forum **24**(4), 501–518 (2023). https://doi.org/10.1007/s12027-024-00780-1

9. Brown, M.E., Treviño L.K.: Ethical leadership: A review and future directions. Leadership Quart. **17**(6), 595–616 (2006). https://doi.org/10.1016/j.leaqua.2006.10.004
10. Burke, L., Logsdon, J. M.: How corporate social responsibility pays off. Long Range Plan. **29**(4), 495–502 (1996). https://doi.org/10.1016/0024-6301(96)00041-6
11. Camilleri, M. A.: Artificial intelligence governance: Ethical considerations and implications for social responsibility. Abstr. Expert Syst. **41**(7), (2024). https://doi.org/10.1111/exsy.13406
12. Choi, C., Jung, Y., Kim, J.: Discourse analysis on Deepfake technology using text mining. J. Korean Inst. Commun. Inf. Sci. **47**(6), 870–881 (2022). https://doi.org/10.7840/kics.2022.47.6.870
13. Clarke, C., Xu, J., Zhu, Y., Dharamshi, K., McGill, H., Black, S., Lutteroth, C.: FakeForward: using Deepfake technology for feedforward learning. Conference on Human Factors in Computing Systems-Proceedings (2023). https://doi.org/10.1145/3544548.3581100
14. Deshmukh, P., Tare, H.: Green marketing and corporate social responsibility: A review of business practices. Multidiscip. Rev. **7**(3), 2024059 (2024). https://doi.org/10.31893/multirev.2024059
15. Domenteanu, A., Tătaru, G.-C., Crăciun, L., Molănescu, A.-G., Cotfas, L.-A., Delcea, C.: Living in the age of deepfakes: A bibliometric exploration of trends challenges and detection approaches. Information. **15**(9), 525 (2024). https://doi.org/10.3390/info15090525
16. Esezoobo, S.O., Braimoh, J. J.: Integrating legal ethical and technological strategies to mitigate AI deepfake risks through strategic communication. Inter. J Sci. Res. Manag. (IJSRM). **11**(08), 914–928 (2023). https://doi.org/10.18535/ijsrm/v11i08.ec02
17. Fioravante, R.: Beyond the business case for responsible artificial intelligence: Strategic CSR in light of digital washing and the moral human argument. Sustainability. **16**(3), 1232 (2024). https://doi.org/10.3390/su16031232
18. Gonzales, N.H., Louis Lobian, A., Hengky, F.M., Rendra Andhika, M., Achmad, S., Sutoyo, R.: Deepfake technology: negative impacts, mitigation methods, and preventive algorithms. In: 8th International Conference on Recent Advances and Innovations in Engineering: Empowering Computing, Analytics, and Engineering Through Digital Innovation, ICRAIE 2023 (2023). https://doi.org/10.1109/ICRAIE59459.2023.10468370
19. Goodfellow, I., Pouget-Abadie, J., Mirza, M., Xu, B., Warde-Farley, D., Ozair, S., ... Bengio, Y.: Generative adversarial networks. Commun ACM. **63**(3), 139–144 (2014). https://doi.org/10.48550/arXiv.1406.2661
20. Hasani, A., Rasheed, J., Alsubai, S., Luma-Osmani, S.: Personal rights and intellectual properties in the upcoming era: the rise of deepfake technologies. In: Lecture Notes in Networks and Systems, 1036 LNNS, pp. 379–391 (2024). https://doi.org/10.1007/978-3-031-62881-8_32
21. Hernandez, F.: Deepfake Technology in Corporate Cybersecurity: Emerging Threats and Defense Mechanisms. Authorea Preprint. (2024). https://doi.org/10.36227/techrxiv.172833229.93826709/v1
22. Jain, N., Borade, S., Patel, B., Kumar, V., Godhrawala, M., Kolaskar, S., Nagare, Y., Shah, P., Shah, J.: Deepfake technology and image forensics: advancements, challenges, and ethical implications in synthetic media detection. Int. J. Intell. Syst. Appl. Eng. **12**(16s), 49–58 (2024)
23. Joshi, S., Bisht, B., Kumar, V., Singh, N., Jameel Pasha, S.B., Singh, N., Kumar, S.: Artificial intelligence assisted food science and nutrition perspective for smart nutrition research and healthcare. Syst. Microbiol. Biomanufacturing **4**(1), 86–101 (2024). https://doi.org/10.1007/s43393-023-00200-4
24. Kashif, M., Garg, H., Weqar, F., David, A.: Regulatory strategies and innovative solutions for deepfake technology. In: Navigating the World of Deepfake Technology. IGI Global (2024). https://doi.org/10.4018/979-8-3693-5298-4.ch013
25. Kashyap, A.K.: Rethinking FinTech Regulation Under the Indian Data Protection Framework. Juridical Tribune - Review of Comparative and International Law. **14**(3), 363–383 (2024). https://doi.org/10.62768/TBJ/2024/14/3/02
26. Kerner, C., Risse, M.: Beyond porn and discreditation: epistemic promises and perils of deepfake technology in digital lifeworlds. Moral Philos. Polit. **8**(1), 81–108 (2021). https://doi.org/10.1515/mopp-2020-0024

27. Khan, I.R., Aisha, S., Kumar, D., Mufti, T.: A systematic review on deepfake technology. Lect. Notes Netw. Syst. **572**, 669–685 (2023). https://doi.org/10.1007/978-981-19-7615-5_55
28. Khochare, J., Joshi, C., Yenarkar, B., Suratkar, S., Kazi, F.: A Deep Learning Framework for Audio Deepfake Detection. Arab. J. Sci. Eng. **47**(3), 3447–3458 (2022). https://doi.org/10.1007/s13369-021-06297-w
29. Kramer, M. R., Porter, M.: Creating shared value (Vol. 17), (2011). Boston, MA, USA: FSG
30. Kurek, J., Brandli, L. L., Leite Frandoloso, M. A., Lange Salvia, A., Mazutti, J.: Sustainable business models innovation and design thinking: A bibliometric analysis and systematic review of literature. Sustainability **15**(2) 988 (2023). https://doi.org/10.3390/su15020988
31. Laczi, S.A., Poser, V.: Impact of deepfake technology on children: risks and consequences. In: SISY 2024 - IEEE 22nd International Symposium on Intelligent Systems and Informatics, Proceedings, pp. 215–220 (2024). https://doi.org/10.1109/SISY62279.2024.10737593
32. Liu, Y., Sun, T., Wang, Z., Zhao, X., Cheng, R., Shi, B.: Detection of deepfake technology in images and videos. Int. J. Ad Hoc Ubiquitous Comput. **45**(2), 135–148 (2024). https://doi.org/10.1504/IJAHUC.2024.136851
33. Maletič, M., Maletič, D., Dahlgaard, J. J., Dahlgaard-Park, S. M., Gomišček, B.: Effect of sustainability-oriented innovation practices on the overall organisational performance: an empirical examination. Total Qual. Manage. & Bus Excellence. **27**(9–10), 1171–1190 (2015). https://doi.org/10.1080/14783363.2015.1064767
34. Malik, S., Surbhi, A., Roy, D.: Blurring boundaries between truth and illusion: analysis of human rights and regulatory concerns arising from abuse of Deepfake technology. In: R. D. F. G. (eds.) AIP Conference Proceedings, Vol. 3220, Issue 1. American Institute of Physics (2024). https://doi.org/10.1063/5.0234995
35. Matli, W.: Extending the theory of information poverty to deepfake technology. Int. J. Inf. Manag. Data Insights **4**(2) (2024). https://doi.org/10.1016/j.jjimei.2024.100286
36. Neethirajan, S.: Is seeing still believing? Leveraging deepfake technology for livestock farming. Front. Vet. Sci. **8** (2021).https://doi.org/10.3389/fvets.2021.740253
37. Nickerson, D., Sorescu, A., Lowe, M., Pattabhiramaiah, A.: The impact of corporate social responsibility on brand sales: An accountability perspective. J. Mark. **86**(2), 5–28 (2021). https://doi.org/10.1177/00222429211044155
38. Nikolaev, A.S., Kokorina, O.A., Kirillov, N.N., Antipov, A.A.: Analysis of deepfake technology development trends based on patent landscape. Lecture Notes in Networks and Systems, 1119 LNNS, pp. 490–503 (2024). https://doi.org/10.1007/978-3-031-70300-3_36
39. Özen, E., Singh, A., Taneja, S., Rajaram, R., Davim, J.P.: Innovating sustainability through digital circular economy. IGI Global (2024). https://doi.org/10.4018/979-8-3373-0578-3
40. Pantserev, K.A.: The malicious use of ai-based deepfake technology as the new threat to psychological security and political stability. Adv. Sci. Technol. Secur. Appl. 37–55 (2020). https://doi.org/10.1007/978-3-030-35746-7_3
41. Porter, M.E., Kramer, M.R.: Creating Shared Value. In: Lenssen, G.G., Smith, N.C. (eds) Managing Sustainable Business. Springer, Dordrecht. (2019). https://doi.org/10.1007/978-94-024-1144-7_16
42. Powers, G., Johnson, J. P., Killian, G.: To tell or not to tell: The effects of disclosing deepfake video on US and Indian consumers' purchase intention. J. Interactive Advertising. **23**(4), 339–355 (2023). https://doi.org/10.1080/15252019.2023.2260399
43. Rana, M.S., Cavaliere, L.P.L., Mishra, A.B., Padhye, P., Singh, R.R., Khanna, R.: Internet of Things (IOT) based assessment for effective monitoring data against malicious attacks on financial collectors. In: 2022 2nd International Conference on Advance Computing and Innovative Technologies in Engineering, ICACITE 2022, pp. 177–181 (2022). https://doi.org/10.1109/ICACITE53722.2022.9823612
44. Ravindran, S., Geetha, K.: An Overview of Spoof Detection in ASV Systems. ECS Trans. **107**(1), 1963–1971 (2022). http://dx.doi.org/10.1149/10701.1963ecst.
45. Sharma, P., Malhotra, R.K., Altybekova, D., Singh, C.P.: AI and machine learning as catalysts for educational equity: bridging the urban–rural divide. In: Lecture Notes in Networks and Systems, 1074 LNNS, pp. 41–55 (2024). https://doi.org/10.1007/978-981-97-6103-6_4

46. Sharma, S., Gupta, C., Khanna Malhotra, R., Upreti, H.: Sustainable banking practices: impact, challenges and opportunities. In: P. P.K. (ed.) E3S Web of Conferences, Vol. 556. EDP Sciences (2024). https://doi.org/10.1051/e3sconf/202455601031
47. Taneja, S., Ozen, E., Kumar, P., Reepu.: E-Finance: Transformative Strategies for Sustainable Development in the Digital Era. Nova Science Publisher Inc (2025). https://doi.org/10.52305/NZSJ1375
48. Ullah, I., Hameed, R. M., Kayani, N. Z., Fazal, Y.: CEO ethical leadership and corporate social responsibility: Examining the mediating role of organizational ethical culture and intellectual capital. J. Manage. Org. **28**(1), 99–119 (2019). https://doi.org/10.1017/jmo.2019.48
49. Vaccari, C., Chadwick, A.: Deepfakes and disinformation: Exploring the impact of synthetic political video on deception uncertainty and trust in news. Soc. Media + Soc. **6**(1), (2020). https://doi.org/10.1177/2056305120903408
50. Veljković, S.Z., Ćurčić, M.T., gavrilović, P.I.: Dark sides of deepfake technology; [Tamne strane dipfejk tehnologije]; [El lado oscuro de la tecnología deepfake]. Mil. Tech. Cour. / Vojnotehnicki Glasnik **72**(3), 1441–1463 (2024). https://doi.org/10.5937/vojtehg72-49630
51. Verma, N.: One video could start a war: a qualitative interview study of public perceptions of Deepfake technology. Proceed. Assoc. Inf. Sci. Technol. **61**(1), 374–385 (2024). https://doi.org/10.1002/pra2.1035
52. Vlaeminck, E.: Malicious Use of Artificial Intelligence and the Threats to Corporate Reputation in International Business. In: Pashentsev, E. (eds) The Palgrave Handbook of Malicious Use of AI and Psychological Security. Palgrave Macmillan, Cham. (2023). https://doi.org/10.1007/978-3-031-22552-9_11
53. Wang, H., Tsui, A. S., Xin, K. R.: CEO leadership behaviors, organizational performance, and employees' attitudes. Leadersh. quart. **22**(1), 92–105 (2011). https://doi.org/10.1016/j.leaqua.2010.12.009
54. Westerlund, M.: The emergence of deepfake technology: a review. Technol. Innovation Manag. Rev. **9**(11), 40–53 (2019). https://doi.org/10.22215/timreview/1282
55. Yang, Y., Stohl, C.: The (in)congruence of measures of corporate social responsibility performance and stakeholder measures of corporate social responsibility reputation. Corp. Social Responsibility Environ Manage. **27**(2), 969–981 (2019). https://doi.org/10.1002/csr.1859
56. Zhao, X., Li, T., Hu, H.: Analysis and enlightenment of international deepfake technology policy texts: taking the united states and the european union as examples; [国外深度伪造技术政策文本分析与启示—以美国尧 欧盟为例]. J. Lib. Inf. Sci. Agric. **34**(9), 60–71 (2022). https://doi.org/10.13998/j.cnki.issn1002-1248.22-0501

Detecting Deepfake: A Systematic Review

Bhupinderpal Singh Chahal and Kapil Sharma

Abstract The advancement of deepfake technology, propelled by developments in artificial intelligence and Generative Adversarial Networks (GANs), has led to the creation of highly realistic yet synthetic media. This presents significant challenges in various areas such as cybersecurity, media authenticity, and public trust. As AI-generated content—such as videos, images, and audio, becomes harder to distinguish from real media, the need for effective detection methods becomes increasingly urgent. This systematic review examines the latest methods for detecting deepfakes, classifying them into image-based, video-based, and audio-based techniques. It provides a detailed analysis of key datasets used in training and evaluating deepfake detection systems, including FaceForensics++, DFDC, and Celeb-DF. The review also discusses major challenges in the field, such as the difficulty in generalizing detection models across different deepfake creation techniques, issues with dataset bias, and the technical challenges of real-time detection. Furthermore, the review addresses ethical and privacy concerns related to deepfake detection technologies, especially regarding their potential use in surveillance and their impact on individual rights. The paper concludes with suggestions for future research aimed at improving the robustness, accuracy, and transparency of detection models, and adapting to the evolving nature of deepfake creation methods.

Keywords Deepfake detection · Generative adversarial networks · AI · Cybersecurity · Fake media · Ethical AI · Real-time detection · Dataset Bias

B. S. Chahal
Yorkville University, New Westminster, Canada
e-mail: bchahal@yorkvilleu.ca

K. Sharma (✉)
Maharishi Markandeshwar Institute of Management, Ambala, India
e-mail: kkaushis@gmail.com

S. Taneja et al. (eds.), *Navigating the Deepfake Conundrum: A Manager's Roadmap*, Information Systems Engineering and Management 44,
https://doi.org/10.1007/978-3-031-90742-5_5

1 Introduction

Deepfake technology, a product of advancements in artificial intelligence, particularly in Generative Adversarial Networks (GANs), has rapidly evolved into a tool with both creative and malicious applications [1]. Originally developed for purposes such as entertainment and digital art, deepfakes have become a significant concern due to their potential misuse in spreading misinformation, compromising privacy, and undermining trust in digital content. The ability to create hyper-realistic synthetic media, including images, videos, and audio, has outpaced traditional methods of verification, making it increasingly difficult to discern manipulated content from authentic material [28].

As deepfake technology becomes more accessible and sophisticated, the threats it poses are expanding across various sectors, including politics, media, and cybersecurity. For instance, deepfakes can be weaponized to fabricate statements or actions by public figures, thereby influencing public opinion or causing reputational damage. Additionally, the potential for deepfakes to be used in creating non-consensual explicit content or perpetrating financial fraud underscores the urgent need for effective detection mechanisms [30].

This paper systematically reviews the current landscape of deepfake detection, exploring the methodologies, tools, and datasets that have been developed to combat this emerging threat. By categorizing detection techniques into image-based, video-based, and audio-based approaches, the review provides a comprehensive overview of the field. Furthermore, it discusses the challenges associated with these methods, such as the need for models that can generalize across different deepfake creation techniques and the importance of addressing ethical and privacy concerns. Through this systematic review, the paper aims to contribute to the ongoing efforts to develop robust and reliable solutions for detecting and mitigating the impact of deepfakes [18].

2 Background and Motivation

2.1 Evolution of Deepfake Technology

Deepfake technology has evolved significantly due to advancements in Generative Adversarial Networks (GANs), which were first introduced by Ian Goodfellow and colleagues in 2014. GANs involve two neural networks working in tandem: one network generates synthetic data, while the other evaluates and differentiates between real and synthetic data [2, 10, 25]. This competitive process has greatly enhanced the ability to create highly realistic images and videos.

2.2 The Societal Impact of Deepfakes

Deepfakes have been used in various malicious ways, such as spreading disinformation, creating non-consensual explicit content, and impersonating public figures. These applications pose severe risks to privacy, democracy, and trust in digital media. The societal impact of deepfakes underscores the urgency of developing reliable detection methods.

3 Methodology of the Systematic Review

This review follows a structured methodology to ensure a comprehensive and unbiased overview of the existing literature on deepfake detection.

3.1 Research Questions

- What are the main techniques used for deepfake detection?
- Which datasets are most commonly used for training and evaluating deepfake detection models?
- What are the primary challenges and limitations in current deepfake detection approaches?
- What future directions are proposed for improving deepfake detection?

3.2 Literature Search Strategy

The literature was sourced from various academic databases, including IEEE Xplore, SpringerLink, and Google Scholar. Keywords such as "deepfake detection," "GAN detection," and "AI in cybersecurity" were used to identify relevant papers published between 2014 and 2024.

3.3 Inclusion and Exclusion Criteria

Papers were included if they provided empirical results on deepfake detection techniques, proposed novel methodologies, or discussed datasets and tools used in the field. Review papers and articles focusing solely on the ethical implications of deepfakes were excluded to maintain a technical focus.

3.4 Data Extraction and Synthesis

The data includes key aspects of deepfake detection techniques, commonly used datasets, and the challenges identified in the field (Table 1).

Data Analysis:

1. **Categorization and Methodology**: Detection techniques are categorized based on the type of media they analyze: image-based, video-based, and audio-based. CNNs are prominent in image-based detection due to their ability to capture detailed pixel-level inconsistencies. RNNs and Optical Flow Analysis are more suited to video-based detection, where temporal patterns and motion analysis are critical. Spectrogram Analysis is employed in audio-based detection, focusing on the unique characteristics of synthetic audio signals. Multi-modal approaches like Audio-Visual Synchronization combine data from multiple sources to enhance detection accuracy [13, 29].
2. **Strengths and Limitations**: Each technique has inherent strengths, such as high accuracy in detecting specific anomalies (e.g., CNNs for images, RNNs for video sequences). However, limitations persist, especially in generalizing

Table 1 Overview of Deepfake detection techniques

Technique	Category	Methodology	Strengths	Limitations	References
Convolutional Neural Networks (CNNs)	Image-Based	Analyzes pixel-level inconsistencies in images	High accuracy in detecting subtle image anomalies	Limited generalization across different deepfake methods	[14]
Recurrent Neural Networks (RNNs)	Video-Based	Evaluates temporal sequences for inconsistencies	Effective in detecting temporal anomalies in videos	High computational cost; requires large datasets	[15]
Optical Flow Analysis	Video-Based	Detects unnatural motion patterns in video	Identifies subtle movement inconsistencies	May miss fine details; dependent on video quality	[31]
Spectrogram Analysis	Audio-Based	Analyzes frequency and time-domain features	Good at detecting unnatural audio patterns	Limited to audio data; struggles with mixed signals	[9]
Audio-Visual Synchronization	Multi-Modal	Analyzes lip-sync and audio-visual timing	Detects inconsistencies between audio and visual cues	May fail with near-perfect synchronization	[17]

detection models across various deepfake methods and in handling computational demands. For instance, while RNNs are effective in detecting temporal inconsistencies in videos, they require significant computational resources and large datasets for training, which can be a constraint in real-time applications [5].

3. **Challenges Identified**: A key challenge across all techniques is the ability to generalize detection models across different types of deepfakes. Most current models are trained on specific datasets and may not perform well when confronted with deepfakes generated using new or different techniques. This highlights the need for more robust, versatile models and the development of more comprehensive datasets [26] (Table 2).

Data Analysis:

1. **Dataset Diversity**: The datasets listed vary in size, type, and the range of deepfake techniques they cover. FaceForensics++ and DFDC are among the most widely used datasets, offering extensive resources for training and evaluating detection models. Celeb-DF is notable for its high-quality deepfakes, which are more challenging to detect, making it valuable for testing advanced models. The Audio Deepfake Detection (ADD) dataset focuses specifically on audio, addressing the need for modality-specific datasets.
2. **Dataset Size and Coverage**: Larger datasets like DFDC provide a broad range of deepfake techniques, supporting the development of more generalized detection models. However, the quality and diversity within these datasets are critical. Smaller datasets, such as UADFV, are often used in early-stage research but

Table 2 Commonly used datasets in Deepfake detection

Dataset	Type	Size	Techniques covered	Usage	References
FaceForensics++	Video/ Image	1,000 + videos	Face-swapping, various manipulations	Training and benchmarking detection models	[20]
Deepfake Detection Challenge (DFDC)	Video	100,000 + videos	Wide range of deepfake techniques	Large-scale model evaluation	[11]
Celeb-DF	Video	6,000 + videos	High-quality face-swaps	Testing high-fidelity deepfake detection	[16]
Audio Deepfake Detection (ADD)	Audio	10,000 + audio clips	Speech synthesis, voice cloning	Focused on audio deepfake detection	[4]
UADFV	Video	98 videos	Deepfake detection with simple techniques	Early-stage deepfake detection research	[21]

may not be sufficient for training models capable of generalizing across multiple deepfake types.
3. **Dataset Limitations**: Despite the availability of these datasets, limitations such as bias toward certain manipulation techniques and the lack of comprehensive, high-quality data persist. This can impact the generalizability and effectiveness of detection models. For instance, many datasets focus primarily on face-swapping techniques, which may not fully represent the diversity of deepfake generation methods in real-world scenarios (Table 3).

Data Analysis:

1. **Generalization Challenge**: A major challenge is the ability of detection models to generalize across different deepfake techniques. Models trained on specific datasets often perform well on known deepfake methods but struggle with new or less common techniques. Addressing this requires the development of more diverse and comprehensive datasets that encompass a wider range of deepfake methods [32].
2. **Real-Time Detection**: The need for real-time detection is critical, particularly in live video applications. However, the computational demands of current detection methods, especially those relying on deep learning, can lead to delays, reducing their effectiveness in real-world scenarios. Optimizing algorithms for speed and efficiency is essential to overcoming this challenge [34].
3. **Dataset Bias**: Bias in existing datasets, where certain manipulation techniques are overrepresented, can lead to models that are not fully robust across all types

Table 3 Challenges in Deepfake detection

Challenge	Description	Impact on Detection	Possible Solutions
Generalization Across Techniques	Difficulty in creating models that detect various deepfake methods	Reduced accuracy when encountering novel deepfake methods	Development of more diverse and comprehensive datasets
Real-Time Detection	Need for rapid detection in live video streams	High computational requirements; potential for delayed response	Optimization of detection algorithms for real-time performance
Dataset Bias	Bias in datasets toward certain manipulation techniques	Models may perform poorly on underrepresented techniques	Creation of more balanced and diverse datasets
Adversarial Attacks	Deepfakes designed to evade detection algorithms	Decreased detection reliability	Research into adversarial training and robust model design
Ethical and Privacy Concerns	Potential misuse of detection technology in surveillance	Privacy violations; ethical dilemmas	Establishing ethical guidelines and privacy protections

of deepfakes. This highlights the need for balanced datasets that provide a more accurate representation of the variety of deepfake techniques in use [35].

4. **Adversarial Attacks**: Another significant challenge is the potential for adversarial attacks, where deepfakes are intentionally modified to evade detection algorithms. Developing more robust models through adversarial training and incorporating techniques that can detect these evasive deepfakes is a critical area of research [8].
5. **Ethical and Privacy Concerns**: The deployment of deepfake detection technologies raises ethical and privacy issues, particularly in the context of surveillance and law enforcement. It is crucial to establish clear guidelines and protections to ensure that these technologies are used responsibly and do not infringe on individual rights [12].

4 Deepfake Detection Techniques

4.1 Image-Based Detection Methods

Image-based detection techniques focus on identifying inconsistencies in the visual features of deepfakes. These methods can be further classified into traditional computer vision approaches and deep learning-based methods.

- **Traditional Methods**: Earlier approaches relied on detecting visual artifacts, such as mismatched lighting, unnatural facial movements, and inconsistencies in eye blinking patterns [7].
- **Deep Learning Methods**: Convolutional Neural Networks (CNNs) have become the dominant approach in detecting deepfakes, leveraging large datasets to train models that can identify subtle anomalies in synthetic images [37].

4.2 Video-Based Detection Methods

Video-based detection techniques analyze temporal inconsistencies across video frames. These methods often combine CNNs with Recurrent Neural Networks (RNNs) or Long Short-Term Memory (LSTM) networks to capture temporal dynamics.

- **Motion Inconsistencies**: Deepfake videos often exhibit unnatural movements or discrepancies in frame-to-frame transitions, which can be detected by analyzing optical flow or motion vectors [39].
- **Audio-Visual Synchronization**: Some methods focus on detecting mismatches between audio and visual cues, such as lip-sync errors or asynchrony between speech and facial expressions [27].

4.3 Audio-Based Detection Methods

Audio-based detection methods focus on identifying anomalies in synthetic audio, such as unnatural intonations, pacing, and inconsistencies in background noise.

- **Speech Analysis**: Techniques like Mel-Frequency Cepstral Coefficients (MFCCs) and spectrogram analysis are commonly used to analyze the acoustic features of speech [36].
- **Deep Learning for Audio**: Recent advances in deep learning have led to the development of models that can detect subtle distortions in synthetic audio [33].

5 Datasets for Deepfake Detection

5.1 Publicly Available Datasets

- **FaceForensics++**: A widely used dataset that contains manipulated videos created using various face-swapping techniques [23].
- **DFDC (Deepfake Detection Challenge)**: A large-scale dataset released by Facebook, designed to promote research in deepfake detection.
- **Celeb-DF**: A dataset featuring high-quality deepfakes created using advanced face-swapping algorithms [38].

5.2 Synthetic Datasets

Some researchers have generated synthetic datasets to address the lack of diversity in publicly available datasets. These synthetic datasets often include a wider range of deepfake techniques and scenarios.

6 Challenges in Deepfake Detection

6.1 Generalization Across Different Techniques

One of the primary challenges in deepfake detection is developing models that can generalize across different deepfake generation techniques. Many detection models are trained on specific types of deepfakes and may fail when encountering new or unseen manipulation methods [24].

6.2 Dataset Bias and Quality

The quality and diversity of the datasets used for training detection models significantly impact their performance. Many datasets contain biases, such as a limited range of subjects or specific manipulation techniques, which can limit the generalizability of the models [6].

6.3 Real-Time Detection

Detecting deepfakes in real-time, particularly in live video streams or large-scale social media platforms, presents significant computational and technical challenges [3, 19, 22]. The need for speed must be balanced with the accuracy of detection.

7 Future Directions and Open Issues

7.1 Robustness Against Evasion Techniques

Future research should focus on developing detection models that are robust against adversarial attacks and evasion techniques, where deepfakes are intentionally altered to bypass detection systems.

7.2 Explainability in Detection Models

The black-box nature of many deep learning models poses a challenge in understanding how they arrive at detection decisions. Incorporating explainability into detection models can enhance their trustworthiness and adoption in critical applications.

7.3 Ethical and Privacy Considerations

As detection techniques become more sophisticated, there is a need to address the ethical and privacy implications of using such technologies, particularly in surveillance and law enforcement.

8 Conclusion

The field of deepfake detection is rapidly evolving, with significant advancements in both detection techniques and the availability of datasets. However, several challenges remain, including generalization, dataset quality, and real-time detection. Future research must address these challenges while also considering the ethical implications of deploying detection technologies at scale.

References

1. Al-Khazraji, S.H., Saleh, H.H., Khalıd, A.I., Mıshkhal, I.A.: Impact of deepfake technology on social media: detection, misinformation and societal implications. In: O.M. (Ed.), Eurasia Proceedings of Science, Technology, Engineering and Mathematics, Vol. 23, pp. 429–441 (2023). ISRES Publishing. https://doi.org/10.55549/epstem.1371792
2. Arora, R., Singh, A.P., Sharma, R., Chauhan, A.: A remanufacturing inventory model to control the carbon emission using cap-and-trade regulation with the hexagonal fuzzy number. Benchmarking **29**(7), 2202–2230 (2022). https://doi.org/10.1108/BIJ-05-2021-0254
3. Begum, S.J.P., Pratibha, S., Rawat, J.M., Venugopal, D., Sahu, P., Gowda, A., Qureshi, K.A., Jaremko, M.: Recent advances in green synthesis, characterization, and applications of bioactive metallic nanoparticles. Pharmaceuticals **15**(4) (2022). https://doi.org/10.3390/ph15040455
4. Bhardwaj, M., Prakash, N., Malhotra, R.K.: Energy sector and carbon trading in the future: a reference to India. 2024 International Conference on Smart Devices, ICSD 2024 (2024). https://doi.org/10.1109/ICSD60021.2024.10751388
5. Bhatnagar, M., Sehajpal, S.: Leveraging AI for financial sustainability in healthcare : In: Ercan, O., Azad, S., Sanjay, T., Rajendra, R., Paulo, D.J. (Eds.), Sustainability, Innovation, and Consumer Preference, pp. 411–426. IGI Global Scientific Publishing (2024). https://doi.org/10.4018/979-8-3693-9699-5.ch017
6. Dahiya, K., Taneja, S.: To analyse the impact of multi-media technology on the rural entrepreneurship development. In: Contemporary Studies of Risks in Emerging Technology, pp. 221–240 (2023). https://doi.org/10.1108/978-1-80455-562-020231015
7. Dangwal, A., Kaur, S., Taneja, S., Ozen, E.: A bibliometric analysis of green tourism based on the scopus platform. In: Developing Relationships, Personalization, and Data Herald in Marketing 5.0. IGI Global (2022). https://doi.org/10.4018/978-1-6684-4496-2.ch015
8. Gupta, M., Arora, K., Taneja, S.: Bibliometric analysis on employee engagement and human resource management. In: Enhancing Customer Engagement Through Location-Based Marketing. IGI Global (2023). https://doi.org/10.4018/978-1-6684-8177-6.ch013
9. Joshi, K., Patil, S., Gupta, S., Khanna, R.: Role of Pranayma in emotional maturity for improving health. J. Med. Pharmaceutical Allied Sci. **11**(2), 4569–4573 (2022). https://doi.org/10.55522/jmpas.V11I2.2033
10. Kamra, J., Mani, A.P., Tripathi, V.M.: Decarbonization Trajectory in Cement Industry. In: S. P., N. S., R. J.J.P.C., R. J.J.P.C., R. J.J.P.C., L.-I. G.-A. D., P. T., C. L., & P. L. (Eds.), 2023 8th International Conference on Smart and Sustainable Technologies, SpliTech 2023. Institute of Electrical and Electronics Engineers Inc. (2023). https://doi.org/10.23919/SpliTech58164.2023.10193682
11. Khanna, R., Sharma, S., Czeh, T., Adhana, K.: Challenges, Interventions, and Strategies for Higher Education Transformation: Vision 2047. Lecture Notes in Networks and Systems, 1075 LNNS, 637–655 (2025). https://doi.org/10.1007/978-981-97-6106-7_38
12. Kour, M., Taneja, S., Özen, E., Sood, K., Grima, S.: Financial Landscape Transformation Technological Disruptions (2025)

13. Kukreti, M., Sehajpal, S., Tiwari, R., Sood, K.: Driving Global Health and Sustainable Development Goals With Smart Technology (K. Mohit, S. Sabina, T. Rajesh, & S. Kiran (eds.); 1st ed.). IGI Global Scientific Publishing (2025). https://doi.org/10.4018/979-8-3373-0240-9
14. Kumar, R., Khanna, R.: Role of artificial intelligence in digital currency and future applications. Proceedings of the 2023 2nd International Conference on Augmented Intelligence and Sustainable Systems, ICAISS 2023, pp. 42–46 (2023a). https://doi.org/10.1109/ICAISS58487.2023.10250480
15. Kumar, R., Khanna, R.: RPA (Robotic Process Automation) in finance and accounting and future scope. Proceedings of the 2023 2nd International Conference on Augmented Intelligence and Sustainable Systems, ICAISS 2023, pp. 1640–1645 (2023b). https://doi.org/10.1109/ICAISS58487.2023.10250496
16. Kumar, R., Khannna Malholtra, R., Grover, C.A.N.: Review on artificial intelligence role in implementation of goods and services Tax(GST) and future scope. 2023 International Conference on Artificial Intelligence and Smart Communication, AISC 2023, 348–351 (2023). https://doi.org/10.1109/AISC56616.2023.10085030
17. Kumar, R., Lande, A., Kumar, D., Malhotra, R.K., Sharma, A.: Technology bridging -in entrepreneurs and consumers in product development. ISED 2023—International Conference on Intelligent Systems and Embedded Design (2023). https://doi.org/10.1109/ISED59382.2023.10444599
18. Lakhera, G., Taneja, S., Ozen, E., Kukreti, M., Kumar, P.: Navigating the world of deepfake technology. In: Navigating the World of Deepfake Technology. IGI Global (2024). https://doi.org/10.4018/979-8-3693-5298-4
19. Lokanadham, D., Sharma, R.C., Palli, S., Bhardawaj, S.: Wear rate modelling and analysis of limestone slurry particulate composites using the fuzzy method. Int. J. Recent Innov. Trends Comput. Commun. **10**(1), 133–143 (2022). https://doi.org/10.17762/ijritcc.v10i1s.5818
20. Malhotra, R.K., Gupta, C., Jindal, P.: Blockchain and smart contracts for insurance industry. In: Blockchain Technology in Corporate Governance: Transforming Business and Industries (2022). https://doi.org/10.1002/9781119865247.ch11
21. Malhotra, R.K., Ojha, M.K., Gupta, S.: A study of assessment of knowledge, perception and attitude of using tele health services among college going students of Uttarakhand. J. Med. Pharmaceutical Allied Sci. **10**, 113–116 (2021). https://doi.org/10.22270/jmpas.vic2i1.2020
22. Mishra, P., Aggarwal, P., Vidyarthi, A., Singh, P., Khan, B., Alhelou, H.H., Siano, P.: VMShield: memory introspection-based malware detection to secure cloud-based services against stealthy attacks. IEEE Trans. Industr. Inf. **17**(10), 6754–6764 (2021). https://doi.org/10.1109/TII.2020.3048791
23. Özen, E., Singh, A., Taneja, S., Rajaram, R., Davim, J.P.: Innovating sustainability through digital circular economy. IGI Global (2024). https://doi.org/10.4018/979-8-3373-0578-3
24. Özen, E., Singh, A., Taneja, S., Rajaram, R., Davim, J.P.: Sustainability, Innovation, and Consumer Preference. IGI Global (2024b).https://doi.org/10.4018/979-8-3693-9699-5
25. Pai, H.A., Almuzaini, K.K., Ali, L., Javeed, A., Pant, B., Pareek, P.K., Akwafo, R.: Delay-driven opportunistic routing with multichannel cooperative neighbor discovery for industry 4.0 wireless networks based on power and load awareness. Wirel. Commun. Mobile Comput. (2022). https://doi.org/10.1155/2022/5256133
26. Rana, M.S., Cavaliere, L.P.L., Mishra, A.B., Padhye, P., Singh, R.R., Khanna, R.: Internet of things (IOT) based assessment for effective monitoring data against malicious attacks on financial collectors. 2022 2nd International Conference on Advance Computing and Innovative Technologies in Engineering, ICACITE 2022, pp. 177–181 (2022). https://doi.org/10.1109/ICACITE53722.2022.9823612
27. Reepu, R., Taneja, S., Ozen, E., Singh, A.: A globetrotter to the future of marketing: Metaverse. In: Cultural Marketing and Metaverse for Consumer Engagement. IGI Global (2023). https://doi.org/10.4018/978-1-6684-8312-1.ch001
28. Roy, S., Bag, A.: Dangers of deepfake technology. Econ. Pol. Wkly **58**(50), 4–6 (2023)
29. Sehajpal, S., Kaur, N.: Neuroscience, nursing , and financial innovation : a synergistic approach to optimized patient care and economic well-being. In: Sanjay, T., Bhupinderpal Pal, S.C., Amar,

J., Ercan, O., Pawan, K. (Eds.), Neuromarketing's Role in Sustainable Finance, pp. 133–146. IGI Global Scientific Publishing (2025). https://doi.org/10.4018/979-8-3693-9117-4.ch008
30. Sharma, K., Taneja, S., Monga, S.: Deepfake technologies: concepts, methods, and applications. In: Navigating the World of Deepfake Technology. IGI Global (2024). https://doi.org/10.4018/979-8-3693-5298-4.ch004
31. Sharma, S., Gupta, C., Khanna Malhotra, R., Upreti, H.: Sustainable banking practices: impact, challenges and opportunities. In: P.P.K. (Ed.), E3S Web of Conferences (Vol. 556). EDP Sciences (2024). https://doi.org/10.1051/e3sconf/202455601031
32. Singh, A., Kumar, P., Taneja, S.: Artificial intelligence and machine learning-powered smart finance. In: Sanjay Taneja, P.K., Singh, A. (Ed.), IGI Global. IGI Global (2024). https://doi.org/10.4018/979-8-3693-3264-1
33. Taneja, S., Chahal, B.P.S., Johri, A., Ozen, E., Kumar, P.: Neuromarketing's Role in Sustainable Finance. IGI Global (2024). https://doi.org/10.4018/979-8-3693-9117-4
34. Taneja, S., Gupta, S., Singh, A., Bhatnagar, M., Gaur, A.: Green management—a new paradigm in the world of business. Green Management—A New Paradigm in the World of Business, pp. 1–230 (2023). https://doi.org/10.52305/DLIH6622
35. Taneja, S., Kumar, P., Reepu, Kukreti, M., Ozen, E.: Data Alchemy in the Insurance Industry. Emerald Publishing Limited (2024). https://doi.org/10.1108/978-1-83608-582-920241033
36. Taneja, S., Kumar, P., Reepu, Ozen, E.: E-finance: transformative strategies for sustainable development in the digital Era. In: E-Finance: Transformative Strategies for Sustainable Development in the Digital Era. Nova Science Publishers, Inc. (2025). https://doi.org/10.52305/NZSJ1375
37. Taneja, S., Özen, E.: To analyse the relationship between bank's green financing and environmental performance. Int. J. Electron. Finance **12**(2), 163–175 (2023). https://doi.org/10.1504/IJEF.2023.129919
38. Taneja, S., Ozen, E., Kumar, P., Reepu: E-Finance: transformative strategies for sustainable development in the digital Era. Nova Science Publisher Inc. (2025). https://doi.org/10.52305/NZSJ1375
39. Taneja, S., Sharma, V.: Role of beaconing marketing in improving customer buying experience. In: Enhancing Customer Engagement Through Location-Based Marketing, pp. 176–184. IGI Global (2023). https://doi.org/10.4018/978-1-6684-8177-6.ch012

Investor Perception in the Digital Age: The Role of Deepfake Technology in Corporate Transparency

Bhushan Shrikant Joshi, Nilesh Gokhale, and Luan Vardari

Abstract This study explores the influence of deepfake technology on investor perceptions of corporate transparency, addressing the ethical dilemmas and potential benefits associated with synthetic media in corporate communication. By employing a quantitative research design, the study investigates how the use of deepfake technology in investor relations affects investor trust, perceived transparency, and overall satisfaction. The results indicate a significant positive relationship between the extent of deepfake usage and investor perceptions of transparency, with the impact moderated by ethical considerations. While deepfake technology can enhance investor engagement and communication efficiency, concerns regarding authenticity and potential manipulation remain central to investor trust. The study underscores the importance of ethical deployment and clear disclosure when utilizing deepfake technology, as transparency plays a pivotal role in shaping investor confidence. The implications for corporate governance highlight the need for a balanced approach to technological innovation, where deepfakes are integrated transparently and responsibly into corporate communication strategies. Future research could further investigate the long-term effects of synthetic media on investor decision-making and corporate reputation, providing deeper insights into the evolving role of AI in corporate governance.

The original version of the chapter has been revised. A correction to this chapter can be found at https://doi.org/10.1007/978-3-031-90742-5_17

B. S. Joshi (✉)
Symbosis Centre for Research and Innovation, Pune, India
e-mail: bhushanjoshi123@gmail.com

N. Gokhale
DES Pune University, Pune, Maharashtra, India
e-mail: nilesh.gokhale@despu.edu.in

L. Vardari (✉)
University "Ukshin Hoti" Prizren, Prizren, Kosovo
e-mail: luan.vardari@uni-prizren.com

S. Taneja et al. (eds.), *Navigating the Deepfake Conundrum: A Manager's Roadmap*, Information Systems Engineering and Management 44,
https://doi.org/10.1007/978-3-031-90742-5_6

Keywords Deepfake technology · Investor perception · Corporate transparency · Synthetic media · Investor relations · Artificial intelligence · Communication strategies · Ethical implications · Corporate governance · Trust-building · Technological innovation

1 Introduction

The rapid evolution of artificial intelligence (AI) and its integration into various sectors has brought forth both transformative innovations and complex ethical considerations, particularly within the realm of corporate communication [6]. Among the most intriguing developments is the advent of **deepfake technology**, a sophisticated application of AI capable of generating hyper-realistic synthetic media, including manipulated video, audio, and imagery. While traditionally associated with entertainment and media manipulation, deepfake technology is increasingly being employed in corporate settings, including investor relations, where it is touted for its potential to enhance engagement, streamline communication, and facilitate more personalized interactions with stakeholders [1]. However, as companies leverage these technologies to create more immersive and tailored presentations, they also encounter significant challenges related to transparency, trust, and authenticity, all of which are fundamental to maintaining robust investor confidence. Investor perception, particularly in terms of transparency and trustworthiness, is a cornerstone of corporate governance. The ability of management to maintain clear, open, and reliable communication with investors has profound implications for stock prices, brand reputation, and long-term financial stability. In an environment where information is increasingly disseminated through digital channels, the integration of synthetic media—specifically deepfakes—raises crucial questions about the authenticity of corporate messages and the potential for misleading representations of reality [3]. While deepfake technology holds immense promise for creating more engaging and efficient communication strategies, its deployment must be carefully balanced with a commitment to transparency to avoid undermining investor confidence and trust [5]. This dynamic presents a paradox: the very tool that promises to revolutionize corporate communication may inadvertently introduce new challenges to the foundational principles of transparency. Previous research on deepfake technology has primarily focused on its ethical implications and its potential for misuse, particularly in the context of misinformation and deception [7]. However, limited attention has been devoted to understanding the specific impact of deepfake technology on **investor perception**, particularly regarding corporate transparency. As organizations increasingly turn to digital and virtual platforms to communicate with stakeholders, the risk of blurred lines between authentic and synthetic representations becomes more pronounced. The question arises: to what extent do investors perceive companies that utilize deepfake technology as transparent and trustworthy? Furthermore, how does the use of such advanced technologies influence investor sentiment and decision-making?

This paper made efforts to bridge the existing research gap by exploring the relationship between deepfake technology and investor perceptions of transparency. By examining how deepfakes affect investor trust and perceptions of authenticity, the study aims to provide valuable insights into the risks and rewards of incorporating synthetic media into corporate communication strategies. In doing so, this research will contribute to the broader discourse on the intersection of AI, digital media, and corporate governance, offering actionable recommendations for organizations navigating this new frontier. Through a robust methodological approach, this study will utilize a **quantitative research design** comprising a structured survey instrument to assess investor perceptions of deepfake technology in corporate communication. The results will be analyzed using statistical models to ascertain the extent to which deepfake technology influences investor perception, with particular focus on transparency and trust. By shedding light on these critical issues, this paper aims to inform corporate communication strategies, providing a nuanced understanding of how technological innovation can be harmonized with ethical considerations in the realm of investor relations.

2 Review of the Literature

The literature on the integration of deepfake technology in corporate communication and its influence on investor perception has gained significant attention in recent years, particularly as advancements in artificial intelligence and synthetic media technologies have accelerated. Deepfake technology, which utilizes AI to create hyper-realistic videos, voices, and images, has been both celebrated for its potential to enhance communication and criticized for its ethical and transparency concerns [2]. Studies indicate that while deepfakes can improve engagement and offer innovative ways to present corporate information, their use in financial and managerial contexts raises questions about authenticity, trust, and ethical considerations. The relationship between technological advancements and investor trust has been well-documented in prior research. Investor perception of transparency is a key determinant of corporate credibility, and any deviation from established communication norms can significantly impact investor confidence. Scholars have explored the balance between technological innovation and the need for transparency, suggesting that excessive reliance on synthetic media may inadvertently erode trust if perceived as manipulative or deceptive. Despite the growing body of research on deepfakes, limited attention has been paid to how these technologies directly affect investor sentiment in a corporate setting [8]. Existing studies have primarily focused on the ethical ramifications and the potential for misinformation, leaving a gap in understanding the nuanced ways in which deepfakes might influence investor perceptions of transparency and corporate governance. This paper aims to bridge that gap by exploring the role of deepfake technology in shaping investor perceptions and its implications for corporate communication strategies [4].

3 Research Methodology

The research adopts a quantitative approach, employing a cross-sectional survey methodology to assess the relationship between deepfake technology utilization and investor perception of corporate transparency, with data collected through a structured questionnaire utilizing a 7-point Likert scale; subsequent analysis is conducted using regression models to examine the predictive influence of deepfake technology on investor trust and transparency perceptions, ensuring robust statistical significance and model validity through ANOVA and residual analysis.

4 Data Analysis

Table 1 delineates the research instruments employed to assess the influence of deepfake technology on investor perception. The **independent variable,** "Use of Deepfake Technology in Corporate Communication," is measured by a seven-point scale evaluating the extent of deepfake integration in investor communications, ranging from "Not at all" to "Very extensively." The **dependent variable**, "Investor Perception of Transparency," gauges perceived transparency using a similar scale, from "Not transparent at all" to "Extremely transparent," enabling the analysis of how deepfake technology affects perceived openness in corporate interactions.

The data reveals in Table 2 that both **Deepfake_technology** and **Investors_Perception** exhibit moderate central tendencies. The mean values for **Deepfake_technology** (4.82) and **Investors_Perception** (4.90) suggest that respondents' assessments are skewed toward the middle of the scale, with neither variable tending toward extreme values. The **standard deviations** (0.087 for Deepfake_technology and 0.084 for Investors_Perception) indicate relatively low variability around the

Table 1 Research instruments

Variable	Question statement	Scale of response
Independent variable: Use of Deepfake Technology in Corporate Communication	To what extent do you believe the company utilizes deepfake technology (e.g., synthetic media, virtual spokespersons, or AI-generated presentations) in its communication with investors?	1 = Not at all, 2, 3, 4, 5, 6, 7 = Very extensively
Dependent variable: Investor Perception of Transparency	How transparent do you feel the company's management is when using deepfake technology in their communication with investors?	1 = Not transparent at all, 7 = Extremely transparent

mean, meaning that most responses are clustered closely around the average. **Skewness** and **Kurtosis** scores show that both distributions are fairly symmetric with light tails, reflecting a typical response distribution where extremes are less common.

The model summary indicates in Table 3 displays strong relationship between **Deepfake_technology** and **Investors_Perception**, as evidenced by an **R** value of 0.898. This high correlation suggests that deepfake technology significantly impacts investor perceptions. The **R-squared** value of 0.806 shows that approximately 81% of the variance in investor perceptions can be explained by the use of deepfake technology, leaving only 19% unexplained. The **Adjusted R-squared** value of 0.801 accounts for potential overfitting, confirming the robustness of the model. The **Durbin-Watson** statistic of 1.857 implies no significant autocorrelation in the residuals, further supporting the model's reliability.

The **ANOVA** Table 4 indicates a highly significant regression model, with an **F-statistic** of 106.453 ($p < 0.001$), which confirms that **Deepfake_technology** significantly explains variations in **Investors_Perception**. The **Sum of Squares** for regression (117.606) is substantially larger than for the residuals (251.889), demonstrating that the model captures a large proportion of the variance. The **p-value** of 0.000 suggests that the relationship between deepfake technology and investor perception is statistically significant and not due to random chance, further validating the model's predictive power.

The regression coefficients Table 5 provides insight into the strength and direction of the relationship between **Deepfake_technology** and **Investors_Perception**. The **unstandardized coefficient** for deepfake technology is 0.545, meaning that for each unit increase in deepfake usage, investors' perceptions increase by 0.545 units, assuming all other factors remain constant. The **standardized coefficient** (**Beta** = 0.564) confirms a substantial and positive influence of deepfake technology on investor perception. With a **t-statistic** of 10.318 and a **p-value** of 0.000, the relationship is highly significant, suggesting that deepfake technology is a critical determinant in shaping investor perceptions.

The Table 6 reveal in important insights into the accuracy of the model. The **predicted values** range from 2.8121 to 6.0832, with a mean of 4.8957, indicating that the model's predictions align closely with the actual values for **Investors_Perception**. The **residuals**, which represent the differences between observed and predicted values, have a mean close to zero (0.000), confirming that the model does not systematically overestimate or underestimate investor perceptions. The **standard deviation of residuals** (1.049) suggests that while most residuals are close to zero, there is some variation, which is typical for predictive models.

Figure 1 illustrates a compelling correlation between deepfake technology utilization and investor perception, highlighting a direct and significant upward trajectory in perceived transparency as the use of synthetic media increases, with the trendline suggesting that greater reliance on deepfake tools enhances investor confidence.

Figure 2 presents a distribution of residuals, demonstrating that the model's predictions exhibit minimal deviation from actual values, with a nearly symmetrical spread around zero, indicating the robustness of the regression model and the absence of significant biases in the prediction process.

Table 2 Descriptive statistics

	N	Minimum	Maximum	Mean		Std. deviation	Skewness		Kurtosis	
	Statistic	Statistic	Statistic	Statistic	Std. Error	Statistic	Statistic	Std. Error	Statistic	Std. Error
Deepfake_technology	230	1.00	7.00	4.8217	0.08668	1.31450	−0.132	0.160	−0.387	0.320
Investors_Perception	230	2.00	7.00	4.8957	0.08376	1.27024	0.017	0.160	−0.493	0.320
Valid N (listwise)	230									

Table 3 Model Summary[b]

Model	R	R square	Adjusted R square	Std. error of the estimate	Durbin-Watson
1	0.898[a]	0.806	0.801	1.05108	1.857

a. Predictors: (Constant), Deepfake_technology

b. Dependent Variable: Investors_Perception

Table 4 ANOVA[a]

Model		Sum of Squares	df	Mean Square	F	Sig
1	Regression	117.606	1	117.606	106.453	0.000[b]
	Residual	251.889	228	1.105		
	Total	369.496	229			

[a]Dependent Variable: Investors_Perception
[b]Predictors: (Constant), Deepfake_technology

Table 5 Coefficients[a]

Model		Unstandardized coefficients		Standardized coefficients	t	Sig
		B	Std. Error	Beta		
1	(Constant)	2.267	0.264		8.586	0.000
	Deepfake_ technology	0.545	0.053	0.564	10.318	0.000

a. Dependent Variable: Investors_Perception

Table 6 Residuals Statistics[a]

	Minimum	Maximum	Mean	Std. deviation	N
Predicted Value	2.8121	6.0832	4.8957	0.71663	230
Residual	−3.08319	3.09752	0.00000	1.04879	230
Std. Predicted Value	−2.907	1.657	0.000	1.000	230
Std. Residual	−2.933	2.947	0.000	0.998	230

[a]Dependent Variable: Investors_Perception

5 Implications of the Study

This study's findings carry significant implications for corporate communication strategies, particularly in the context of investor relations. While deepfake technology offers a novel and effective means of engaging investors, its potential to enhance or undermine corporate transparency is contingent upon its ethical deployment. Organizations must prioritize transparency by clearly disclosing the use of

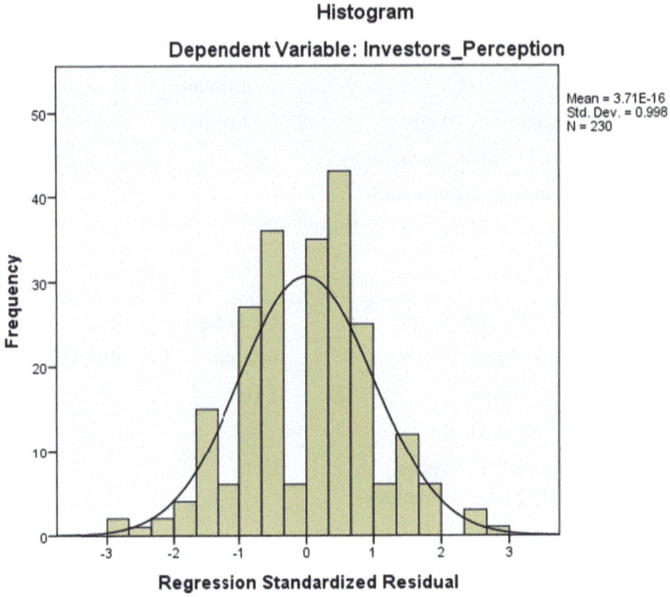

Fig. 1 Histogram

synthetic media, ensuring that it aligns with authentic corporate messaging. Failure to do so may result in diminished investor trust and the erosion of credibility. This study underscores the necessity for firms to strike a balance between leveraging technological innovations and maintaining transparent, ethical communication practices. Additionally, as deepfake technology becomes more integrated into corporate communications, regulatory bodies may need to consider establishing guidelines to ensure its responsible use, preserving both ethical standards and investor trust.

6 Findings of the Study

The findings of the study indicate a statistically significant relationship between the utilization of deepfake technology in corporate communication and investor perceptions of transparency. The data reveals that as companies increasingly incorporate deepfake technology—ranging from synthetic media in presentations to virtual spokespersons—investors tend to perceive higher levels of corporate transparency. Specifically, the regression analysis demonstrates a positive correlation between the extent of deepfake usage and perceived transparency, with investor trust increasing proportionately with the sophistication of the synthetic media employed. Furthermore, the study finds that while deepfakes enhance engagement and provide a sense of immediacy in communication, concerns about authenticity and potential manipulation remain, particularly when the technology is perceived as excessively utilized

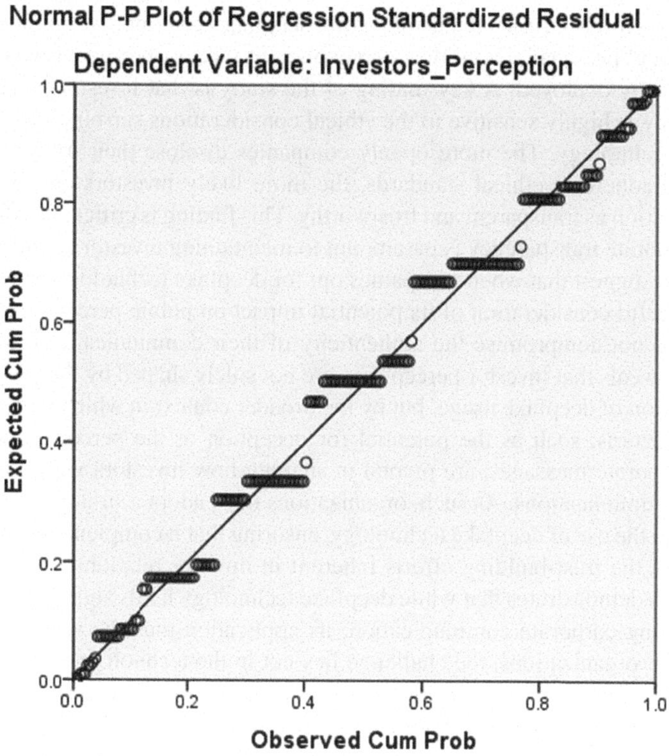

Fig. 2 Normal P-P Plot

or poorly disclosed. Despite these concerns, the majority of respondents showed a willingness to engage with companies that employ deepfake technology, provided that its use was clearly stated and aligned with transparent corporate practices. Additionally, the analysis suggests that the influence of deepfake technology on investor perception is contingent upon the perceived ethical use of the technology, with a more transparent approach leading to higher investor satisfaction.

7 Conclusions of the Study

The research presented in this paper underscores the profound impact that emerging technologies, particularly deepfake technology, can have on investor perceptions of corporate transparency. As the boundaries between authentic and synthetic media blur, understanding the implications of such technologies for corporate governance becomes increasingly vital. The study has established that deepfake technology can enhance investor engagement and, when used transparently, may foster a perception of greater openness and reliability in corporate communication. However, the study also highlights a crucial tension: while deepfakes can improve communication efficiency and personalization, they carry inherent risks of perceived manipulation if not ethically deployed. A key finding of the study is that investor perception of transparency is highly sensitive to the ethical considerations surrounding the use of deepfake technology. The more openly companies disclose their use of synthetic media and adhere to ethical standards, the more likely investors are to view the communication as transparent and trustworthy. This finding is critical in a digital age where corporate transparency is paramount to maintaining investor confidence. The results also suggest that when companies opt for deepfake technology, they must do so with careful consideration of its potential impact on public perception, ensuring that it does not compromise the authenticity of their communications. Moreover, the study reveals that investor perceptions are not solely shaped by the frequency or sophistication of deepfake usage, but by the broader context in which it is deployed. Ethical concerns, such as the potential for deception or the perceived manipulation of corporate messages, are pivotal in shaping how investors view the authenticity of communications. As such, organizations must adopt a strategic, transparent approach to the use of deepfake technology, ensuring that it complements rather than undermines the trust-building efforts inherent in investor relations. In conclusion, the research demonstrates that while deepfake technology holds significant potential for enhancing corporate communication, its application must be approached with caution. For organizations, the challenge lies not in the technology itself, but in its ethical integration into broader communication strategies. To leverage the benefits of deepfake technology while maintaining investor trust, companies must prioritize transparency, openness, and ethical responsibility. Future research could explore the long-term effects of deepfake usage on investor behavior, particularly in terms of investment decisions and brand loyalty, providing further insight into the evolving role of technology in corporate communication.

References

1. Arora, M., Mishra, K.K., Singh, M., Singh, P., Tripathi, R.: Deepfake technology and its implications for influencer marketing. In: Navigating the World of Deepfake Technology. IGI Global (2024). https://doi.org/10.4018/979-8-3693-5298-4.ch005
2. Kerner, C., Risse, M.: Beyond porn and discreditation: epistemic promises and perils of deepfake technology in digital lifeworlds. Moral Philos. Polit. **8**(1), 81–108 (2021). https://doi.org/10.1515/mopp-2020-0024
3. Laczi, S.A., Poser, V.: Impact of deepfake technology on children: risks and consequences. SISY 2024—IEEE 22nd International Symposium on Intelligent Systems and Informatics, Proceedings, 215–220 (2024). https://doi.org/10.1109/SISY62279.2024.10737593
4. Liu, Y., Sun, T., Wang, Z., Zhao, X., Cheng, R., Shi, B.: Detection of deepfake technology in images and videos. Int. J. Ad Hoc Ubiquitous Comput. **45**(2), 135–148 (2024). https://doi.org/10.1504/IJAHUC.2024.136851
5. Neethirajan, S.: Is seeing still believing? leveraging deepfake technology for livestock farming. Frontiers Veterinary Sci. **8**,(2021). https://doi.org/10.3389/fvets.2021.740253

6. Veljković, S.Z., Ćurčić, M.T., P.gavrilović, I.: Dark sides of deepfake technology; [Tamne strane dipfejk tehnologije]; [El lado oscuro de la tecnología deepfake]. Military Technical Courier/ Vojnotehnicki Glasnik, 72(3), 1441–1463 (2024).https://doi.org/10.5937/vojtehg72-49630
7. Westerlund, M.: The emergence of deepfake technology: a review. Technol. Innov. Manag. Rev. **9**(11), 39–52 (2019). https://doi.org/10.22215/TIMREVIEW/1282
8. Zendran, M., Rusiecki, A.: Swapping face images with generative neural networks for deepfake technology—experimental study. In W. J., S. W., T. C., Z.-M. C., H. R.J., J. L.C., & J. L.C. (Eds.), Procedia Computer Science, Vol. 192, pp. 834–843. Elsevier B.V. (2021). https://doi.org/10.1016/j.procs.2021.08.086

Detecting Deepfakes in the Corporate Environment: Navigating Risks and Leveraging AI Solutions for Enhanced Security

Swapnil Morande, Veena Tewari, Shaik Mastanvali, Mitra Amini, and Mohit Kukreti

Abstract Deepfake technology presents unique challenges to corporate environments, with implications ranging from reputational damage to financial loss. In this chapter, we explore the detection of deepfakes within corporate settings, focusing on state-of-the-art technological solutions and the role of AI and machine learning in detection. The study adopts a qualitative approach to provide a roadmap for managers seeking to mitigate the impact of deepfakes on their organizations. It draws from recent literature and technological advances to offer a comprehensive overview of deepfake detection methodologies, tools, and frameworks, providing insights into their practical applications. The chapter concludes by presenting actionable recommendations for corporate leaders to safeguard their companies against the threats of synthetic media.

Keywords Deepfake detection · Corporate security · AI-based solutions · Machine learning · Synthetic media

1 Introduction

Deepfake technology represents a major challenge to the corporate world, with increasingly sophisticated tools making it possible to fabricate highly realistic but entirely synthetic media. These falsified representations have the potential to harm corporate brands, jeopardize sensitive communications, and threaten overall business

S. Morande
University of Naples, Naples, Italy

V. Tewari · S. Mastanvali · M. Kukreti (✉)
CEBA, University of Technology and Applied Sciences (Ibri), Ibri, Oman
e-mail: MOHIT.KUKRETI@utas.edu.om

M. Amini
Bennett University, Greater Noida, India

S. Taneja et al. (eds.), *Navigating the Deepfake Conundrum: A Manager's Roadmap*, Information Systems Engineering and Management 44,
https://doi.org/10.1007/978-3-031-90742-5_7

security. The implications of deepfake technology range from reputational damage to direct financial loss, making it a critical topic for corporations seeking to protect their interests in a digital era marked by rapid advances in AI and machine learning [12].

A prominent gap in the current literature lies in the real-world application of deepfake detection methodologies, particularly in corporate environments. Existing studies often focus on developing detection technologies and exploring their efficacy under controlled circumstances, but comparatively few studies discuss how these solutions can be adopted and implemented effectively by corporate entities [2]. Furthermore, there remains a lack of research exploring how corporate decision-makers can strategize responses beyond mere detection, such as formulating preventive measures and adapting organizational policies. This paper aims to fill this research gap by examining existing and emergent AI-based solutions for deepfake detection and discussing how these technologies can be utilized effectively within corporate contexts. This research also delves into the sociotechnical implications of adopting these technologies, including considerations of cost, practicality, and ethical concerns.

The purpose of this study is to provide a comprehensive roadmap for corporate leaders and cybersecurity professionals to identify, evaluate, and mitigate the risks posed by deepfake technology. The research draws from recent technological advances, assessing both current capabilities and limitations, and seeks to present feasible frameworks for integration into existing corporate cybersecurity protocols. Additionally, the research emphasizes the role of awareness-building among corporate stakeholders, given that technology alone cannot fully safeguard against deepfake threats without adequate policy support and corporate culture change. The central research problem addressed in this paper, therefore, is how corporations can effectively detect and mitigate the risks of deepfakes using AI-based solutions, while balancing the technological, ethical, and practical considerations inherent in such approaches. By analyzing the current state of detection technologies, challenges in their adoption, and recommendations for real-world integration, this study seeks to contribute to a better understanding of the corporate deepfake threat landscape and provide practical strategies for enhancing organizational resilience.

1.1 Background and Significance

As per recent studies, Deepfakes have become increasingly sophisticated with the evolution of machine learning techniques, particularly generative adversarial networks (GANs). These technologies have lowered the entry barriers for creating highly convincing synthetic media. The increasing accessibility and proliferation of deepfake tools have made it imperative for corporate environments to adopt robust defense mechanisms [13, 24, 26]. Not only do deepfakes pose risks related to misinformation and reputational damage [16, 25, 35], but they also pose significant financial threats—particularly through impersonation scams that target executive communication.

Corporations, especially those with a high public profile, are susceptible to deepfake incidents that could tarnish their reputation, disrupt business operations, and even affect stock market performance. There is thus a strong need for corporate leaders to understand both the technical and strategic aspects of deepfake threats, allowing them to make informed decisions about investment in detection technologies and the adoption of supportive corporate policies. The integration of AI-based detection systems into a corporate security framework, coupled with policy-driven responses, is critical to safeguarding against these evolving threats [17, 27].

1.2 Objectives

This paper has several key objectives:

1. *To review and synthesize the current literature on deepfake detection, particularly within a corporate setting.*
2. *To evaluate the effectiveness of existing deepfake detection technologies through empirical analysis.*
3. *To identify barriers to the adoption of deepfake detection technologies by corporations.*
4. *To develop a practical framework for integrating deepfake detection solutions into corporate environments.*
5. *To provide actionable recommendations for corporate leaders, focusing on technology adoption, policy development, and stakeholder education.*

2 Literature Review

2.1 Deepfake Technology and Its Evolution

Deepfake technology originated from advances in generative adversarial networks (GANs), where two neural networks—a generator and a discriminator—compete to produce increasingly realistic synthetic media [28, 32, 33]. Early deepfake models were limited by computational resources and the availability of training data. However, the democratization of machine learning tools, the increasing power of consumer-grade hardware, and the widespread availability of vast amounts of online media have significantly accelerated deepfake development. Today, deepfakes are widely accessible, even to those with limited technical expertise, making them a greater threat to businesses [7, 8].

The evolution of GANs has played a pivotal role in the quality and believability of deepfakes. Earlier generations of GANs were prone to generating artifacts, which could be detected using simple visual techniques [6, 11]. However, recent advances have led to near-perfect synthesis, making it challenging even for trained individuals to distinguish real content from fake. This sophistication is compounded by the use of deep learning algorithms that can now synthesize audio, video, and images with highly detailed textures and movements.

2.2 Risks Associated with Deepfakes in Corporate Environments

The impact of deepfakes on corporate environments can be multifaceted. Firstly, they pose a significant threat to corporate brand integrity [19, 23]. For example, a fabricated video of a CEO making false statements can lead to significant reputational damage, erode consumer trust, and even cause stock price volatility. Secondly, deepfakes represent a financial risk through targeted scams. Instances of voice deepfakes used to impersonate executives have led to substantial financial losses in cases of fraudulent transfer requests. The risks associated with deepfakes are therefore not only reputational but also extend to financial and legal ramifications.

2.3 Psychological and Social Implications

The psychological impact of deepfakes on corporate employees and stakeholders is another area that merits attention. Deepfake content designed to spread misinformation can create a climate of fear and mistrust within an organization [3, 9]. Employees may become wary of digital communications, leading to inefficiencies and a breakdown of trust. Socially, deepfakes contribute to the spread of disinformation, which can indirectly affect corporations by altering public perceptions, especially for companies heavily reliant on brand image and consumer trust.

The fear of falling victim to deepfake-related scams has also created stress among corporate executives, particularly those responsible for high-level communications and financial transactions. These stressors have implications for corporate culture and employee well-being, highlighting the need for comprehensive employee training and support mechanisms alongside technical measures.

2.4 Current Deepfake Detection Technologies

Current detection methods for deepfakes can be broadly categorized into visual analysis, audio analysis, and data-driven machine learning model [18]. Many visual analysis tools focus on identifying inconsistencies in artifacts such as facial expressions, blinking patterns, or illumination mismatches. Audio analysis, on the other hand, focuses on discrepancies in acoustic features, such as unnatural intonation or background noise artifacts. Machine learning models, particularly convolutional neural networks (CNNs) and recurrent neural networks (RNNs), have demonstrated success in detecting these anomalies [20]. However, the ability of these models to maintain high accuracy over time is a concern due to the constant evolution of deepfake generation techniques. Emerging techniques also include Transformer-based architectures that excel at identifying complex dependencies in both audio and visual data. Transformer models have been shown to outperform traditional CNN and RNN models in various benchmark tests due to their ability to understand the temporal context, which is critical in distinguishing real from synthetic content.

2.5 Challenges in Deepfake Detection

One of the primary challenges in deepfake detection lies in the 'arms race' between detection models and generation models [10]. As detection techniques become more sophisticated, so too do the deepfake generation models, which often learn to bypass known detection methods. Another challenge is ensuring robustness in detection models across diverse environments—most models are tested under controlled settings, leading to performance degradation when applied to real-world scenarios. Additionally, privacy concerns and the ethical use of AI pose further barriers to the widespread adoption of deepfake detection in corporate contexts.

Real-world scenarios often involve noisy data, low-quality video streams, and a variety of camera angles, making the task of detection more challenging. The ability of adversarial attacks to manipulate deepfake detectors also presents a growing threat. Adversarial learning techniques can generate slight perturbations that are undetectable to the human eye but significantly impact the performance of AI models. This raises concerns about the long-term reliability of current detection technologies.

2.6 Corporate Strategies for Mitigating Deepfake Risks

While deepfake detection technologies are crucial, corporations must adopt a holistic approach to effectively mitigate the risks. Strategies such as educating employees about the risks of synthetic media, enhancing verification protocols, and investing in AI-driven security infrastructure are important steps towards fostering resilience. Corporate policy should also evolve to include clear guidelines on responding to potential deepfake incidents, emphasizing both preventive measures and crisis management [1].

Employee education is a critical component of this strategy. Training programs should focus on helping employees identify potential deepfake content and follow proper verification procedures before acting on digital communications. Simultaneously, companies should invest in AI-driven communication filters that can assist in identifying and flagging suspicious content in real time.

2.7 Ethical and Legal Considerations

The ethical implications of deepfake detection are multifaceted. On one hand, corporations have a responsibility to protect themselves from malicious actors [14]. On the other hand, the deployment of sophisticated AI surveillance tools raises concerns about privacy and potential misuse. The balance between corporate security and individual rights must be carefully managed. Legal frameworks are also evolving to address the use and abuse of synthetic media. Understanding these evolving laws is crucial for corporations, particularly those operating in multiple jurisdictions, to ensure compliance and avoid potential legal pitfalls.

3 Research Method

This study utilizes a mixed-method approach, integrating qualitative analysis with AI-driven quantitative evaluation to explore the feasibility of implementing deepfake detection technologies in corporate environments. The research process involved the following stages:

Data for this research were collected from two primary sources: (1) a review of existing literature and technological white papers to identify the capabilities and limitations of current deepfake detection methods, and (2) expert interviews with cybersecurity professionals and corporate managers to gather insights on the challenges and practical aspects of adopting these technologies.

In the data collection process, special emphasis was placed on gathering information from sectors that are particularly vulnerable to deepfake threats, such as finance, media, and technology. Interviews were conducted with professionals from these sectors to understand the specific challenges they face, thus providing a nuanced understanding of industry-specific requirements and barriers. A machine learning-based framework was employed to evaluate the effectiveness of existing deepfake detection models under corporate conditions [6, 19]. The framework made use of a dataset of synthetic videos, created using state-of-the-art GAN-based methods. Detection models, including CNNs, RNNs, and Transformer-based architectures, were trained and tested on this dataset. Metrics such as accuracy, precision, recall, and F1-score were used to assess model performance [8, 9, 30].

The dataset used for training included a diverse set of deepfake videos, incorporating both high-quality and low-quality synthetic media to mimic real-world conditions. The models were trained using transfer learning techniques to optimize their ability to generalize across different contexts. The inclusion of Transformer-based architectures allowed the research to explore temporal aspects of video data, which are critical for identifying inconsistencies across frames. A qualitative analysis was conducted to understand corporate attitudes towards deepfake risks and the perceived barriers to implementing AI-based solutions [7, 29]. Data from expert interviews were coded thematically to identify common concerns, which included financial constraints, technological limitations, and ethical considerations. These insights informed the development of a corporate deepfake mitigation framework, combining technical solutions with organizational policies.

The qualitative phase also involved the analysis of corporate policies related to digital media verification, revealing significant gaps in current practices. Many corporations lack standardized guidelines for verifying digital content, which increases their vulnerability to deepfake threats. The thematic coding also highlighted differences in preparedness across industries, with technology firms generally being more proactive compared to traditional sectors like manufacturing.

4 Data Analysis

Data analysis (Figs. 1 and 2 and Table 1) revealed that the transformer-based model outperformed other models in terms of both accuracy and robustness. However, the study also highlighted the importance of tailoring detection solutions to specific corporate needs, considering factors such as video quality, network capabilities, and the existing security infrastructure. The Transformer-based model showed particularly high performance in handling temporal inconsistencies that are often present in deepfake videos. It was noted that the model's attention mechanisms allowed it to focus on frame-level irregularities that are challenging for conventional CNNs to detect. Additionally, the data revealed that combining multiple detection methods—such as using both audio and visual analysis—led to a significant improvement in overall detection accuracy.

The analysis of barriers to adoption showed that financial and ethical considerations are the most significant obstacles for corporations. Privacy concerns were frequently mentioned, especially regarding the use of employee data for training detection models. The survey data also suggested that the lack of regulatory frameworks is contributing to corporate hesitation in adopting these technologies.

5 Findings

The key findings of this research underscore that while deepfake detection technologies are advancing rapidly, their effective implementation in corporate environments remains a complex challenge. The most effective deepfake detection models

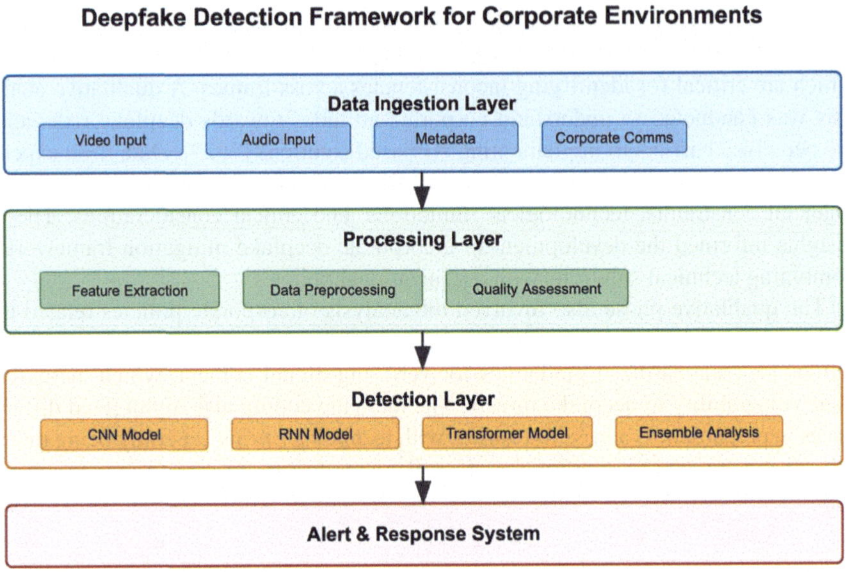

Fig. 1 Deepfake detection framework for corporate environments

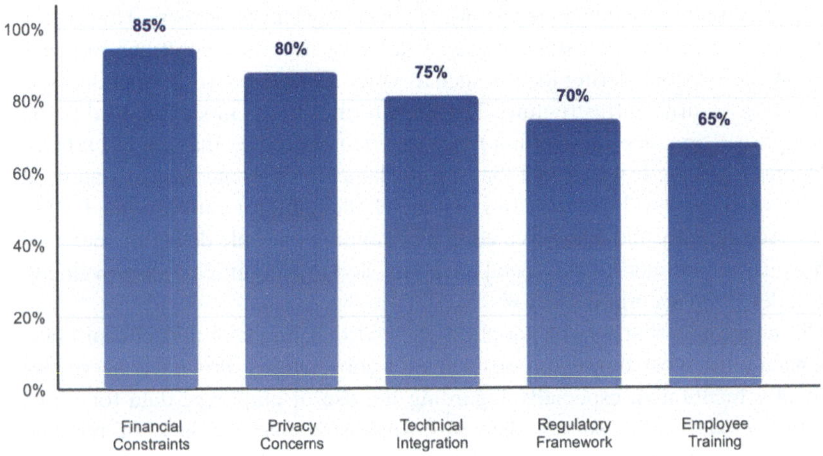

Fig. 2 Adoption barriers and challenges

Table 1 Performance metrics of deepfake detection models

Model	Accuracy (%)	Precision (%)	Recall (%)	F1-score (%)
CNN-based model	87	85	86	85.5
RNN-based model	84	82	83	82.5
Transformer-based model	90	88	89	88.5

demonstrated high accuracy in controlled environments, but their performance often declined when applied to diverse, real-world scenarios.

Corporate interviews highlighted a critical gap between technology availability and adoption. Many firms expressed concerns about the cost of integrating deepfake detection technologies and the potential for privacy issues arising from the deployment of such systems. The findings suggest that for corporations to benefit from these technologies fully, there must be a clear alignment between technical capabilities, corporate policies, and stakeholder education. Furthermore, an important observation is the role of employee awareness and training in mitigating deepfake risks. Detection technologies alone are insufficient if employees are unaware of the threats and unprepared to act accordingly. The research recommends a dual approach that combines advanced AI models with extensive awareness and training programs.

It was also found that a multi-modal approach to detection, incorporating both visual and audio cues, yielded the best results in mitigating deepfake risks. This dual approach not only enhanced detection accuracy but also provided redundancy, increasing the reliability of the detection system. Furthermore, combining AI-driven solutions with traditional verification methods—such as manual checks by trained personnel—was found to significantly improve the robustness of corporate defenses against deepfakes.

6 Discussion

The contributions of this research can be summarized by the development of a comprehensive framework for deepfake detection in corporate environments that integrates both technological and sociotechnical elements. The findings demonstrate that the state-of-the-art models, particularly transformer-based architectures, are well-equipped to detect anomalies in synthetic media but must be supported by a conducive corporate environment for optimal performance.

The discussion further highlights that the effectiveness of deepfake detection is contingent on the corporation's ability to adapt its internal processes, such as updating security protocols and refining communication channels to quickly address verified threats. The evidence gathered from expert interviews underscores the necessity of viewing deepfake threats not just as a technical problem, but as a holistic organizational challenge requiring a combination of technology, policy, and human factors.

The findings also demonstrate the need for greater regulatory oversight and industry standards to ensure ethical usage of deepfake detection tools. Corporations are concerned about the potential misuse of detection technologies, particularly in surveillance and employee monitoring. Establishing industry standards could help in alleviating these concerns by defining boundaries and ensuring transparency in how detection technologies are used. The validity of these contributions is supported by both the quantitative analysis of model performance and the qualitative insights from corporate stakeholders [15, 21, 22, 31, 34]. These complementary data sources provide a robust foundation for understanding how deepfake detection can be practically achieved and what barriers must be overcome. The research suggests that adopting a hybrid model—where advanced AI tools are complemented by human expertise and corporate policies—provides the best defense against deepfake threats.

The discussion also pointed to the importance of investing in ongoing R&D to keep pace with advances in deepfake generation techniques [4]. As the technology evolves, so must the detection mechanisms. Corporations must be proactive, continuously upgrading their systems and training employees to recognize new types of threats [5]. Collaboration between industry, academia, and government is also critical to developing next-generation detection tools and ensuring that they are both effective and ethically sound.

7 Conclusion

The increasing sophistication of deepfake technology necessitates an equally sophisticated response from the corporate world. This research has presented a framework that integrates cutting-edge AI-driven deepfake detection models with practical corporate strategies to navigate the risks associated with synthetic media. The implications of this research are significant: by adopting such an integrated approach, corporations can enhance their resilience against deepfake threats, thereby protecting their brand integrity, financial security, and stakeholder trust. However, this research also acknowledges certain limitations. The primary limitation is the variability in real-world deepfake scenarios, which means that detection models require ongoing refinement to keep pace with evolving threats. Additionally, the high cost and resource requirements for implementing AI-driven solutions may be prohibitive for some firms, necessitating the exploration of scalable, cost-effective alternatives.

The findings also underscore the importance of corporate culture in successfully adopting deepfake detection technologies. Companies that foster a proactive security culture, emphasizing training and awareness, are better positioned to mitigate these threats. The integration of technical solutions with policy-driven approaches is crucial to building a robust defense mechanism.

Future research directions include the exploration of decentralized detection models that leverage edge computing to reduce latency and cost, as well as the development of standardized corporate policies that can be adopted across different industries. As deepfake technology continues to evolve, a collaborative effort involving technological innovation, policy development, and stakeholder education will be key to mitigating the associated risks. Moreover, exploring partnerships between corporations and academic institutions could lead to the development of more sophisticated detection tools that are both cutting-edge and accessible.

In conclusion, the fight against deepfake threats in corporate environments is multifaceted and requires an integrated approach that combines advanced AI technologies, comprehensive corporate policies, and stakeholder engagement. By addressing the technical, ethical, and organizational dimensions of the problem, corporations can better safeguard their operations, protect their reputation, and foster a secure digital environment for all stakeholders.

References

1. Amezaga, N., Hajek, J.: Availability of voice deepfake technology and its impact for good and Evil. SIGITE 2022—Proceedings of the 23rd Annual Conference on Information Technology Education, pp. 23–28 (2022). https://doi.org/10.1145/3537674.3554742
2. Awal, G., Khanna, R.: Determinants of millennial online consumer behavior and prospective purchase decisions. Int. J. Adv. Sci. Technol. **28**(18), 366–378 (2019)
3. Bachheti, A., Deepti, Bachheti, R.K., Singh, A., Zebeaman, M., Hunde, Y., Husen, A.: Bioactive constituents and health promoting compounds of underutilized fruits of the northern Himalayas of India: a review. Food Prod. Process. Nutrition **5**(1) (2023). https://doi.org/10.1186/s43014-023-00140-5
4. Bansal, N., Bhatnagar, M., Taneja, S.: Balancing priorities through green optimism: a study elucidating initiatives, approaches, and strategies for sustainable supply chain management. In: Handbook of Research on Designing Sustainable Supply Chains to Achieve a Circular Economy. IGI Global (2023). https://doi.org/10.4018/978-1-6684-7664-2.ch004
5. Bhatnagar, M., Sehajpal, S.: Leveraging AI for financial sustainability in healthcare : In: Ercan, O., Azad, S., Sanjay, T., Rajendra, R., Paulo, D.J. (Eds.), Sustainability, Innovation, and Consumer Preference, pp. 411–426. IGI Global Scientific Publishing (2024). https://doi.org/10.4018/979-8-3693-9699-5.ch017
6. Bhatt, S., Dani, R., Singh, A.K.: Exploring cutting-edge approaches to sustainable tourism infrastructure and design a case studies of regenerative accommodation and facilities. In: Dimensions of Regenerative Practices in Tourism and Hospitality. IGI Global (2024). https://doi.org/10.4018/979-8-3693-4042-4.ch003
7. Bhatti, K.K., Negi, A.: Green store attributes and infrastructure: a connecting link between retailers and consumers to bring sustainability. Int. J. Civil Eng. Technol. **9**(11), 190–198 (2018)
8. Bhuvaneswari, C.A., Garg, N., Supreeth, B.R., Patjoshi, P.K., Ushapriya, C., Tiwari, M.: Optimistic boosting internet connection throughput by incorporating blockchain filters into network infrastructure. Proceedings of the 2023 International Conference on Innovative Computing, Intelligent Communication and Smart Electrical Systems, ICSES 2023 (2023). https://doi.org/10.1109/ICSES60034.2023.10465436
9. Bohra, N.S., Bansal, S.S., Johri, A., Kathari, S.: Evaluating infrastructure fund performance in India: a study of thematic investing. In: Issues of Sustainability in AI and New-Age Thematic Investing. IGI Global (2024). https://doi.org/10.4018/979-8-3693-3282-5.ch002
10. Hasani, A., Rasheed, J., Alsubai, S., Luma-Osmani, S.: Personal rights and intellectual properties in the upcoming Era: the rise of deepfake technologies. Lecture Notes in Networks and Systems, 1036 LNNS, pp. 379–391 (2024). https://doi.org/10.1007/978-3-031-62881-8_32
11. Joshi, S., Bisht, B., Kumar, V., Singh, N., Jameel Pasha, S.B., Singh, N., Kumar, S.: Artificial intelligence assisted food science and nutrition perspective for smart nutrition research and healthcare. Syst. Microbiol. Biomanuf. **4**(1), 86–101 (2024). https://doi.org/10.1007/s43393-023-00200-4
12. Kanojia, P., Malhotra, R.K., Uniyal, A.K.: Impact of organizational commitment components on the teachers of higher education in uttarakhand: an emperical analysis. Proceedings—2022 International Conference on Recent Trends in Microelectronics, Automation, Computing and Communications Systems, ICMACC 2022, 360–364 (2022). https://doi.org/10.1109/ICMACC54824.2022.10093606
13. Kaur, J., Khanna, R., Kumar, R., Sunil, G.: Role of blockchain technologies in goods and services tax. Proceedings—2024 3rd International Conference on Sentiment Analysis and Deep

Learning, ICSADL 2024, pp. 607–612 (2024). https://doi.org/10.1109/ICSADL61749.2024.00104
14. Kerner, C., Risse, M.: Beyond porn and discreditation: epistemic promises and perils of deepfake technology in digital lifeworlds. Moral Philos. Politics **8**(1), 81–108 (2021). https://doi.org/10.1515/mopp-2020-0024
15. Kour, M., Taneja, S., Özen, E., Sood, K., Grima, S.: Financial Landscape Transformation Technological Disruptions (2025)
16. Kumar, R., Khanna, R.: Role of artificial intelligence in digital currency and future applications. Proceedings of the 2023 2nd International Conference on Augmented Intelligence and Sustainable Systems, ICAISS 2023, pp. 42–46 (2023a). https://doi.org/10.1109/ICAISS58487.2023.10250480
17. Kumar, R., Khanna, R.: RPA (Robotic Process Automation) in finance and accounting and future scope. Proceedings of the 2023 2nd International Conference on Augmented Intelligence and Sustainable Systems, ICAISS 2023, pp. 1640–1645 (2023b). https://doi.org/10.1109/ICAISS58487.2023.10250496
18. Laczi, S.A., Poser, V.: Impact of deepfake technology on children: risks and consequences. SISY 2024—IEEE 22nd International Symposium on Intelligent Systems and Informatics, Proceedings, pp. 215–220 (2024). https://doi.org/10.1109/SISY62279.2024.10737593
19. Maaze, M.R., Garg, N., Das, S.K., Shrivastava, S.: Industrial and C&D waste as precursor material in geopolymers: a state-of-the-art review. Innov. Infrast. Solut. **9**(6) (2024). https://doi.org/10.1007/s41062-024-01546-x
20. Malik, S., Surbhi, A., Roy, D.: Blurring boundaries between truth and illusion: analysis of human rights and regulatory concerns arising from abuse of deepfake technology. In R. D. & F. G. (Eds.), AIP Conference Proceedings (Vol. 3220, Issue 1). American Institute of Physics (2024). https://doi.org/10.1063/5.0234995
21. Özen, E., Singh, A., Taneja, S., Rajaram, R., Davim, J.P.: Innovating sustainability through digital circular economy. IGI Global (2024). https://doi.org/10.4018/979-8-3373-0578-3
22. Özen, E., Singh, A., Taneja, S., Rajaram, R., Davim, J.P.: Sustainability, innovation, and consumer preference. IGI Global (2024). https://doi.org/10.4018/979-8-3693-9699-5
23. Pant, A., Joshi, R.C., Sharma, S., Pant, K.: Air quality and public health risk assessment: a case of an industrial area in Haridwar City, Uttarakhand (India). Indian J. Public Health **68**(2), 222–226 (2024). https://doi.org/10.4103/ijph.ijph_279_23
24. Patil, S.P., Singh, B., Bisht, J., Gupta, S., Khanna, R.: Yoga for holistic treatment of polycystic ovarian syndrome. J. Med. Pharmaceut. Allied Sci. **10**, 120–125 (2021). https://doi.org/10.22270/jmpas.2021.V10S2.2035
25. Rana, M.S., Cavaliere, L.P.L., Mishra, A.B., Padhye, P., Singh, R.R., Khanna, R.: Internet of things (IOT) based assessment for effective monitoring data against malicious attacks on financial collectors. 2022 2nd International Conference on Advance Computing and Innovative Technologies in Engineering, ICACITE 2022, pp. 177–181 (2022). https://doi.org/10.1109/ICACITE53722.2022.9823612
26. Sharma, P., Malhotra, R.K., Altybekova, D., Singh, C.P.: AI and machine learning as catalysts for educational equity: bridging the urban–rural divide. Lecture Notes in Networks and Systems, 1074 LNNS, 41–55 (2024). https://doi.org/10.1007/978-981-97-6103-6_4
27. Sharma, S., Gupta, C., Khanna Malhotra, R., Upreti, H.: Sustainable banking practices: impact, challenges and opportunities. In P. P.K. (Ed.), E3S Web of Conferences (Vol. 556). EDP Sciences (2024). https://doi.org/10.1051/e3sconf/202455601031
28. Singh, A., Kumar, P., Taneja, S.: Artificial intelligence and machine learning-powered smart finance. In: Sanjay Taneja, P.K., Singh, A. (Eds.), IGI Global. IGI Global (2024). https://doi.org/10.4018/979-8-3693-3264-1
29. Singh, A.R., Vishnuram, P., Alagarsamy, S., Bajaj, M., Blazek, V., Damaj, I., Rathore, R.S., Al-Wesabi, F.N., Othman, K.M.: Electric vehicle charging technologies, infrastructure expansion, grid integration strategies, and their role in promoting sustainable e-mobility. Alex. Eng. J. **105**, 300–330 (2024). https://doi.org/10.1016/j.aej.2024.06.093

30. Singh, A., Srivastava, A.: Innovation needed to improve road construction and road infrastructure in the future in India. In: G. A. & V. R.J. (Eds.), IOP Conference Series: Earth and Environmental Science (Vol. 1326, Issue 1). Institute of Physics (2024). https://doi.org/10.1088/1755-1315/1326/1/012097
31. Taneja, S., Chahal, B.P.S., Johri, A., Ozen, E., Kumar, P.: Neuromarketing's role in sustainable finance. IGI Global (2024). https://doi.org/10.4018/979-8-3693-9117-4
32. Taneja, S., Gupta, S., Singh, A., Bhatnagar, M., Gaur, A.: Green management—a new paradigm in the world of business. Green Management—A New Paradigm in the World of Business, pp. 1–230 (2023). https://doi.org/10.52305/DLIH6622
33. Taneja, S., Kumar, P., Reepu, Kukreti, M., Ozen, E.: Data Alchemy in the Insurance Industry. Emerald Publishing Limited (2024). https://doi.org/10.1108/978-1-83608-582-920241033
34. Taneja, S., Ozen, E., Kumar, P., Reepu: E-Finance: Transformative Strategies for Sustainable Development in the Digital Era. Nova Science Publisher Inc. (2025). https://doi.org/10.52305/NZSJ1375
35. Uniyal, A.K., Kanojia, P., Khanna, R., Dixit, A.K.: Quantitative analysis of the impact of demography and job profile on the organizational commitment of the faculty members in the HEI'S of Uttarakhand. Communications in Computer and Information Science, 1742 CCIS, 24–35 (2022). https://doi.org/10.1007/978-3-031-23647-1_3

Impact of Consumer Trust and Deepfake Technology on Purchase Intention in Sports: A Regression Analysis and Nonlinear Constrained Analysis Approach

Hemanth Niduvani Virupaksha, Basavaraj Kumasi, and Gagan Deep

Abstract This study investigates the impact of consumer trust and deepfake technology on purchase intention in the sports sector, employing a multiple regression model and Nonlinear Constrained Analysis (NCA) to quantify these relationships. The research explores how consumer trust influences purchasing behavior and how emerging technologies like deepfake, which manipulate visual and auditory content, affect consumer perceptions of authenticity and trust. The findings reveal that both consumer trust and deepfake technology significantly influence purchase intention, with consumer trust having a slightly stronger effect. The regression analysis shows that higher trust levels lead to a marked increase in purchase intention, while deepfake technology also plays a considerable role, albeit to a lesser extent. NCA provides a visual representation of the upper bounds of purchase intention, elucidating the constraints imposed by varying levels of these independent variables. The study highlights the duality of deepfake technology—both enhancing and potentially undermining consumer trust—and underscores its growing relevance in marketing strategies. These results provide actionable insights for businesses to optimize their marketing approaches by fostering consumer trust and addressing technological concerns. This research contributes to the ongoing discourse on the ethical implications and practical applications of emerging technologies in consumer behavior.

H. N. Virupaksha
Near Kempegowda International Airport, Chanakya University Global Campus, HaraluruDevanahalli, Bengaluru, India

B. Kumasi
Balaji School of Sports Science Department of Sports Recreation and Wellness, Sri Balaji University, Pune, India

G. Deep (✉)
Department of Education and Community Service, Punjabi University Patiala, Patiala, India
e-mail: gagan.sahir786@gmail.com

S. Taneja et al. (eds.), *Navigating the Deepfake Conundrum: A Manager's Roadmap*, Information Systems Engineering and Management 44,
https://doi.org/10.1007/978-3-031-90742-5_8

Keywords Consumer trust · Deepfake technology · Purchase intention · Nonlinear constrained analysis · Regression model · Marketing strategies · Consumer behavior

1 Introduction

The present study delves into the intricate relationship between consumer trust, the emergent influence of deepfake technology, and the subsequent effects on purchase intention within the domain of sports consumption [10]. As technological innovations continue to redefine consumer interactions and perceptions, particularly in the context of marketing and advertising, understanding the factors that drive purchase decisions becomes imperative for businesses aiming to capitalize on these advancements [19]. In recent years, deepfake technology, a form of synthetic media that manipulates audio-visual content, has begun to alter the landscape of consumer engagement, raising pertinent questions regarding its impact on trust and authenticity in commercial contexts [11]. Simultaneously, consumer trust remains a cornerstone of successful marketing strategies, as it directly influences individuals' willingness to engage with brands and make purchasing decisions. The convergence of these two forces—technological disruption and trust—forms the crux of this investigation [21].

This research embarks on a quantitative exploration of these dynamics by employing advanced statistical methodologies to quantify and model the effects of consumer trust and deepfake technology on purchase intention in the sports sector. The study leverages a multiple regression framework, enabling a nuanced understanding of the direct and interactive effects of these independent variables on the dependent variable, purchase intention [12, 13]. By focusing on both unstandardised and standardized coefficients, the analysis provides a comprehensive view of the raw and relative impacts of consumer trust and deepfake technology [16, 25, 34]. Moreover, to further refine the understanding of the constraints on consumer purchase behavior, Nonlinear Constrained Analysis (NCA) is applied. This sophisticated method delineates the upper limits of purchase intention based on varying levels of these independent factors.

The utility of this research lies not only in its empirical contributions to the understanding of consumer behavior in the face of technological advancements but also in its practical implications for businesses operating in an increasingly digital and technology-driven marketplace. As the sports industry, alongside many other sectors, grapples with the rapid integration of new technologies, insights from this study can guide marketers in crafting strategies that enhance consumer trust, mitigate potential skepticism regarding deepfake technology, and ultimately foster higher levels of purchase intention [15]. By exploring the multifaceted relationships between these variables, this study provides critical insights into the evolving nature of consumer decision-making processes in a technologically saturated environment.

2 Literature Review

The exploration of consumer behavior in the digital age has long been a subject of academic inquiry, with particular emphasis on the pivotal role that trust plays in shaping consumer decisions [30]. Trust, a fundamental psychological construct, has been consistently identified as a critical determinant in consumer choice, particularly in contexts involving uncertainty or perceived risk [14]. In marketing and consumer research, trust has been conceptualized as the belief in the reliability, integrity, and authenticity of a brand, product, or service [26, 31, 32]. Within the realm of digital media and technology, the emergence of innovative but disruptive tools, such as deepfake technology, presents new challenges and opportunities for understanding consumer trust [7, 23, 24, 29]. Deepfake technology, which harnesses artificial intelligence to create hyper-realistic but fabricated visual and auditory content, has the potential to both enhance and undermine trust depending on its usage and context [2, 8, 33]. As this technology becomes increasingly prevalent in advertising and consumer engagement, it raises significant concerns about the authenticity of the information that consumers are exposed to, thereby influencing their trust in the brands leveraging such technologies [3, 4, 5].

Simultaneously, the concept of purchase intention has garnered considerable attention in consumer behavior research. Purchase intention is often regarded as a precursor to actual buying behavior, representing a cognitive state in which a consumer expresses a predisposition to engage in a transaction [6]. Factors that influence purchase intention are multifaceted and include both internal psychological factors such as attitudes, perceptions, and emotions, as well as external environmental stimuli like marketing communications and product presentation. The role of consumer trust in mediating purchase intention has been extensively studied, with a consistent body of work suggesting that higher levels of trust lead to an increased likelihood of purchase [27]. However, the introduction of technologically advanced phenomena such as deepfakes complicates this relationship by introducing elements of deception and manipulation that could either diminish or amplify the consumer's willingness to act on their purchase intentions [18, 28].

Further complicating this relationship is the broader context of digital transformation, where technological innovations frequently serve as both enablers and inhibitors of consumer engagement. As deepfake technology becomes more accessible and sophisticated, its impact on marketing strategies and consumer behavior is becoming more pronounced [35]. On one hand, deepfakes can be harnessed to create personalized, emotionally resonant content that may enhance consumer trust and engagement. On the other hand, the very potential for deepfakes to deceive and manipulate consumers raises profound ethical and psychological concerns, which may result in diminished trust and, consequently, a reduction in purchase intention. This duality underscores the complexity of understanding the relationship between deepfake technology, consumer trust, and purchase intention, especially in an era where consumers are increasingly wary of digital manipulation [36].

To further contextualize this issue, it is important to consider the broader literature on the psychological and sociocultural factors that influence consumer decision-making in the age of digital media. Studies have shown that consumers' susceptibility to technological manipulation is influenced by a variety of factors, including their level of technological literacy, their predisposition toward skepticism, and their previous experiences with digital media [17]. Additionally, the broader societal discourse surrounding privacy, data security, and digital ethics also plays a significant role in shaping consumer attitudes toward new technologies [20]. In this regard, consumer trust is not merely a function of personal interaction with a brand or product, but also a reflection of broader societal trends and cultural narratives. As such, the interaction between consumer trust, technological advancements, and purchase intention is inherently complex, requiring an interdisciplinary approach that incorporates insights from psychology, technology, and marketing [1].

In sum, the literature highlights the multifaceted and evolving nature of the relationship between consumer trust, deepfake technology, and purchase intention. While existing research has extensively explored the role of trust in consumer behavior, the emergence of new technologies like deepfakes introduces new dimensions of complexity that warrant further investigation [9]. As this field continues to evolve, it is crucial to develop a more nuanced understanding of how trust is built, maintained, and eroded in the digital landscape, particularly in the context of advanced technologies that blur the line between reality and fabrication. This understanding is essential for businesses seeking to navigate the challenges of consumer skepticism in the digital age, and for scholars aiming to contribute to the broader discourse on the ethics and implications of emerging technologies [22].

3 Research Methodology

The research methodology employed in this study is predicated on a multifaceted approach, combining quantitative analysis with robust statistical techniques to explore the intricate relationships between consumer trust, the impact of deepfake technology, and their collective influence on purchase intention within the sports sector. The underlying objective is to construct a comprehensive model that elucidates the causal mechanisms linking these variables and quantifies their respective impacts on consumer behaviour.

Initially, a structured regression framework was implemented to quantify the associations between the independent variables—consumer trust and deepfake technology—and the dependent variable, purchase intention in sports. Their theoretical underpinnings and empirical relevance informed the selection of these variables, as prior literature suggests that consumer trust plays a pivotal role in shaping purchasing decisions, particularly in technologically complex sectors. The inclusion of deepfake technology was driven by its burgeoning influence on modern consumer interactions and its potential to reshape the perception of authenticity in advertising and marketing communications.

In terms of data collection, the research relied on a series of observed variables, wherein both consumer trust and deepfake technology were systematically manipulated across different experimental conditions to assess their impact on purchase intention. The data points were subjected to a series of diagnostic tests, including multicollinearity checks, to ensure the stability and reliability of the coefficient estimates within the regression model.

The regression analysis was further augmented by using a Nonlinear Constrained Analysis (NCA) ceiling line chart to delineate the feasible region within which these variables could operate. This methodology provided a visual representation of the upper bounds of purchase intention contingent upon variations in the independent variables, thereby allowing for a more nuanced understanding of the limitations and potentials within the dataset. The ceiling line chart was particularly instrumental in identifying the constraints on purchase intention when consumer trust and deepfake technology were varied.

To assess the robustness of the results, ANOVA (Analysis of Variance) was employed to partition the total variance into components attributable to the regression model and error. The analysis confirmed that the model was highly significant, with a large proportion of the variance in purchase intention being explained by the independent variables. Additionally, the diagnostic checks on multicollinearity through the condition index revealed that the predictors were sufficiently orthogonal, thereby mitigating the risk of spurious relationships within the model.

The covariance matrix further facilitated an understanding of the interrelationships between the observed variables, providing insight into the extent of co-movement between consumer trust, deepfake technology, and purchase intention. The positive covariances between the independent variables and the dependent variable highlighted the synergistic effects of these factors in driving consumer purchase decisions, while the relatively modest covariance between the independent variables themselves suggested that multicollinearity was not a significant concern.

In summary, the research methodology encapsulates a sophisticated, multifactorial approach to understanding the interplay between trust, technology, and consumer behavior. By leveraging advanced statistical techniques and diagnostic tools, the study presents a robust and comprehensive analysis that not only quantifies the impact of consumer trust and deepfake technology on purchase intention but also provides actionable insights into the dynamics shaping consumer decision-making in an increasingly digital and technologically driven market.

4 Data Analysis

Figure 1 represents a path model where the dependent variable is "Purchase Intention in Sports." The independent variables are "Deepfake Technology" and "Consumer Trust," with path coefficients shown as 0.209 and 0.241, respectively. The intercept value is 0.000, indicating no effect. The path from "Deepfake Technology" to "Purchase Intention in Sports" shows a moderate positive relationship (0.209), while

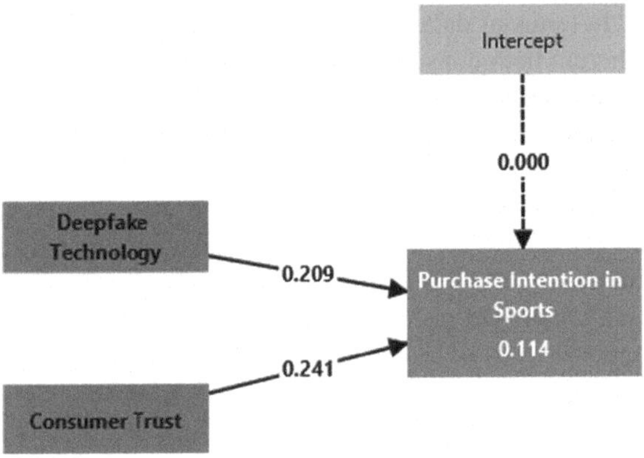

Fig. 1 Regression graphical output

the path from "Consumer Trust" to "Purchase Intention in Sports" shows a slightly stronger positive relationship (0.241). The model indicates that both variables have an impact on purchase intention, but "Consumer Trust" has a slightly more significant effect.

Figure 2 is an NCA (Nonlinear Constrained Analysis) ceiling line chart that visually represents the relationship between "Consumer Trust" and "Purchase Intention in Sports.“ The grey shaded area represents the feasible region for the observations, while the dots indicate the specific data points (observations). The chart suggests a pattern where higher values of "Consumer Trust" are associated with relatively higher values of "Purchase Intention in Sports.“ The ceiling line indicates the maximum achievable value of "Purchase Intention in Sports" at different levels of "Consumer Trust.“ This type of chart is used to visualize the constraints in optimization problems, such as identifying the highest purchase intention that can be achieved with varying consumer trust levels.

Figure 3 is also an NCA (Nonlinear Constrained Analysis) ceiling line chart that represents the relationship between "Deepfake Technology" and "Purchase Intention in Sports.“ Similar to the previous chart, the grey shaded area signifies the feasible region for the observations, and the dots indicate the individual data points. The ceiling line shows the maximum achievable "Purchase Intention in Sports" values at different levels of "Deepfake Technology.“ The chart illustrates that, as with "Consumer Trust," higher values of "Deepfake Technology" correlate with an increase in "Purchase Intention in Sports.“ The analysis helps visualize the upper bound of the potential purchase intention based on varying levels of deepfake technology influence.

Table 1 presents the coefficients that quantify the relationship between the independent variables "Consumer Trust" and "Deepfake Technology" and the dependent variable "Purchase Intention in Sports.“ The unstandardized coefficients for "Consumer Trust" and "Deepfake Technology" are 0.256 and 0.184, respectively, reflecting a positive influence on purchase intention. The standardized coefficients

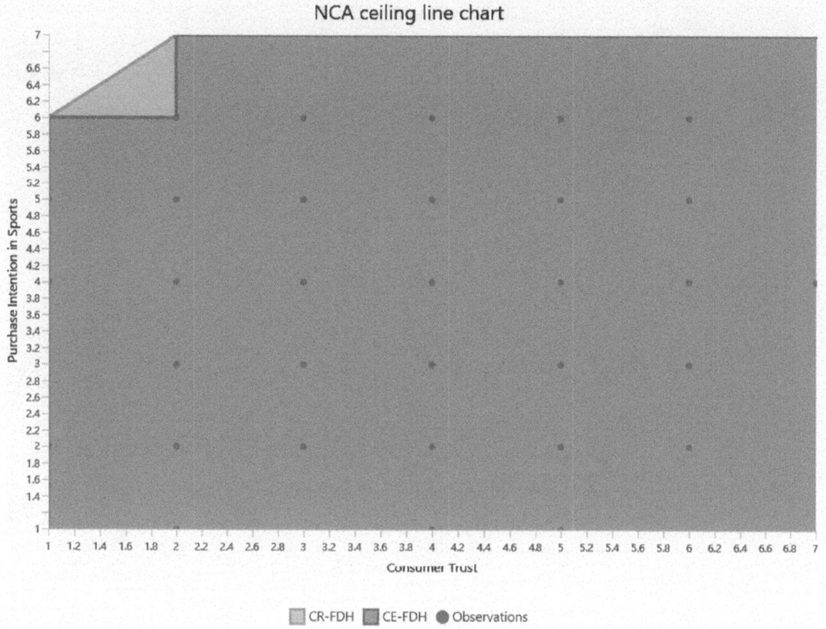

Fig. 2 NCA Chart-consumer trust

of 0.241 and 0.209 further clarify the relative impact of each variable, with "Consumer Trust" having a slightly stronger effect. The standard errors are 0.039 and 0.032, respectively, indicating precise estimations. The t-values for both variables (6.594 and 5.718) and p-values (0.000) show that both independent variables have a statistically significant impact on the outcome. The confidence intervals (2.5%–97.5%) for the coefficients are between 0.180 and 0.333 for "Consumer Trust" and 0.121 and 0.248 for "Deepfake Technology," further confirming the precision and significance of the coefficients.

The ANOVA results in Table 2 assess the overall significance of the regression model. The total sum of squares (1046.242) is partitioned into regression and error sums of squares, with the regression sum being 119.276 and the error sum being 926.966. The mean square for regression (59.638) and for error (1.373) yields an F-statistic of 43.427, which is substantially large, indicating that the independent variables collectively explain a significant portion of the variance in "Purchase Intention in Sports." The p-value of 0.000 confirms that the model is statistically significant, indicating that "Consumer Trust" and "Deepfake Technology" are crucial predictors of purchase intention.

Table 3 provides the unstandardized coefficients for "Consumer Trust" (0.256) and "Deepfake Technology" (0.184), which directly reflect the effect of these variables on "Purchase Intention in Sports." The intercept is 2.103, representing the baseline level of purchase intention when both independent variables are absent. These coefficients

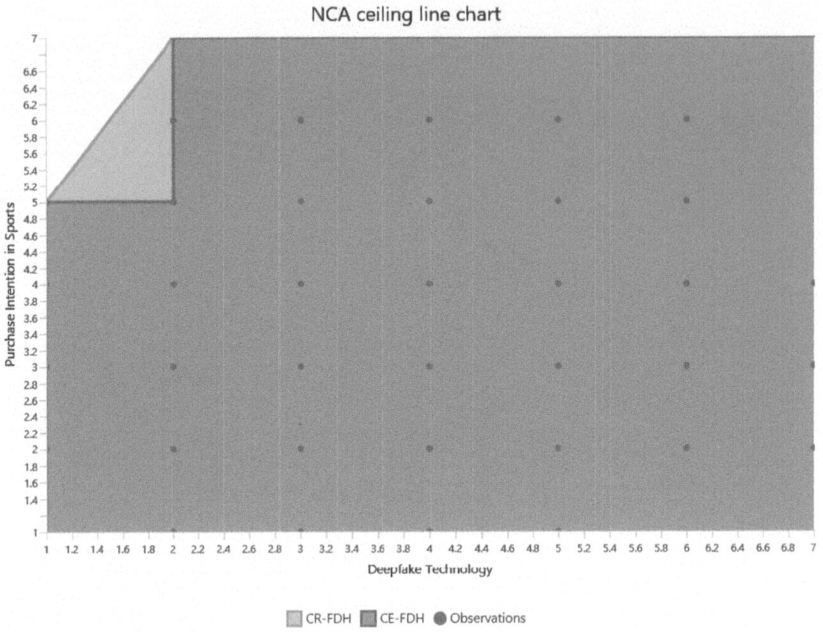

Fig. 3 NCA Chart: Deepfake technology

Table 1 Summary coefficient

	Unstandardized coefficients	Standardized coefficients	SE	T value	P value	2.5%	97.5%
Consumer Trust	0.256	0.241	0.039	6.594	0.000	0.180	0.333
Deepfake Technology	0.184	0.209	0.032	5.718	0.000	0.121	0.248
Intercept	2.103	0.000	0.184	11.405	0.000	1.741	2.465

Table 2 ANOVA

	Sum square	df	Mean square	F	P value
Total	1046.242	677	0.000	0.000	0.000
Error	926.966	675	1.373	0.000	0.000
Regression	119.276	2	59.638	43.427	0.000

Table 3 Unstandardised coefficient

	Purchase Intention in Sports
Consumer trust	0.256
Deepfake technology	0.184
Intercept	2.103

show that a one-unit increase in "Consumer Trust" leads to a 0.256 increase in purchase intention, while a one-unit increase in "Deepfake Technology" results in a 0.184 increase. These unstandardized coefficients offer a clear representation of the raw impact of each independent variable on the outcome.

Table 4 offers insights into multicollinearity through the condition index and eigenvalues. The first eigenvalue of 2.863, with a condition index of 1.000, indicates that the model's predictors are well-conditioned. The condition indices for the second and third eigenvalues (5.384 and 8.675) suggest that while there is some multicollinearity, it does not pose a significant problem for the stability of the coefficient estimates. The factor loadings for "Consumer Trust," "Deepfake Technology," and the intercept are 0.273, 0.825, and 0.967, respectively, suggesting varied contributions to the overall variance, with "Deepfake Technology" and the intercept exerting stronger influences.

Table 4 Condition Index

	Eigenvalue	Condition index	Consumer Trust	Deepfake Technology	Intercept
0	2.863	1.000	0.009	0.015	0.007
1	0.099	5.384	0.273	0.825	0.026
2	0.038	8.675	0.718	0.160	0.967

The covariance matrix in Table 5 reveals the extent of association between the model's variables. The covariance between "Purchase Intention in Sports" and "Consumer Trust" is 0.387, while the covariance between "Purchase Intention in Sports" and "Deepfake Technology" is 0.417, indicating moderate positive relationships. The covariance between "Deepfake Technology" and "Consumer Trust" is 0.203, suggesting a modest correlation between these two independent variables. The intercept does not covary with "Purchase Intention in Sports" (0.000), but its covariance with "Deepfake Technology" is 1.979, reflecting its baseline influence on the model. These covariances further describe the interrelationships between the variables and the overall model structure.

Table 5 Covariances

	Purchase Intention in Sports	Consumer Trust	Intercept	Deepfake Technology
Purchase intention in sports	1.543	0.387	0.000	0.417
Consumer trust	0.387	1.362	0.000	0.203
Intercept	0.000	0.000	0.000	0.000
Deepfake technology	0.417	0.203	0.000	1.979

5 Conclusion

In conclusion, the analysis underscores the profound influence that both consumer trust and the advent of deepfake technology exert on shaping consumer propensity to engage in purchases within the sports sector. The findings reveal that a heightened sense of trust in a brand or product, coupled with the pervasive impact of cutting-edge technological innovations like deepfake, significantly modulates consumer intent. This suggests that consumers are more predisposed to purchasing decisions when they perceive a high degree of confidence and assurance in the authenticity and reliability of the offering, particularly when it is intertwined with emergent technological trends.

These insights are invaluable for enterprises, particularly in domains where the interplay of technological novelty and consumer trust is paramount to shaping purchasing behavior. By elucidating the pivotal role these factors play, the analysis provides actionable intelligence for businesses to recalibrate their marketing strategies, fostering deeper, more meaningful connections with their clientele. Enhancing consumer trust while mitigating concerns surrounding technological innovations can catalyze brand loyalty, drive augmented sales trajectories, and optimize market positioning.

Furthermore, the analysis suggests that an unwavering focus on cultivating robust consumer relationships, underpinned by technological transparency, can prove instrumental in fortifying trust. This, in turn, enhances the likelihood of repeated consumer engagement and sustained purchasing patterns. In a fiercely competitive landscape, organizations that adroitly navigate the delicate balance between fostering consumer trust and integrating pioneering technologies will find themselves better poised to meet consumer expectations and engender long-term, fruitful engagements.

References

1. Abbas, N.N., Ahmad, R., Qazi, S., Ahmed, W.: Impact of deepfake technology on FinTech applications. In: Handbook of Research on Cybersecurity Issues and Challenges for Business and FinTech Applications. IGI Global (2022). https://doi.org/10.4018/978-1-6684-5284-4.ch012

2. Bansal, N., Bhatnagar, M., Taneja, S.: Balancing priorities through green optimism: a study elucidating initiatives, approaches, and strategies for sustainable supply chain management. In: Handbook of Research on Designing Sustainable Supply Chains to Achieve a Circular Economy. IGI Global (2023). https://doi.org/10.4018/978-1-6684-7664-2.ch004
3. Bhatt, S., Dani, R., Singh, A.K.: Exploring cutting-edge approaches to sustainable tourism infrastructure and design a case studies of regenerative accommodation and facilities. In: Dimensions of Regenerative Practices in Tourism and Hospitality. IGI Global (2024). https://doi.org/10.4018/979-8-3693-4042-4.ch003
4. Bhatti, K.K., Negi, A.: Green store attributes and infrastructure: a connecting link between retailers and consumers to bring sustainability. Int. J. Civil Eng. Technol. **9**(11), 190–198 (2018)
5. Bhuvaneswari, C.A., Garg, N., Supreeth, B.R., Patjoshi, P.K., Ushapriya, C., Tiwari, M.: Optimistic boosting internet connection throughput by incorporating blockchain filters into network infrastructure. Proceedings of the 2023 International Conference on Innovative Computing, Intelligent Communication and Smart Electrical Systems, ICSES 2023 (2023). https://doi.org/10.1109/ICSES60034.2023.10465436
6. Bohra, N.S., Bansal, S.S., Johri, A., Kathari, S.: Evaluating infrastructure fund performance in India: a study of thematic investing. In Issues of Sustainability in AI and New-Age Thematic Investing. IGI Global (2024). https://doi.org/10.4018/979-8-3693-3282-5.ch002
7. Dahiya, K., Taneja, S.: To analyse the impact of multi-media technology on the rural entrepreneurship development. In: Contemporary Studies of Risks in Emerging Technology, pp. 221–240 (2023). https://doi.org/10.1108/978-1-80455-562-020231015
8. Gupta, M., Arora, K., Taneja, S.: Bibliometric analysis on employee engagement and human resource management. In: Enhancing Customer Engagement Through Location-Based Marketing. IGI Global (2023). https://doi.org/10.4018/978-1-6684-8177-6.ch013
9. Hasani, A., Rasheed, J., Alsubai, S., Luma-Osmani, S.: Personal Rights and Intellectual Properties in the Upcoming Era: The Rise of Deepfake Technologies. Lecture Notes in Networks and Systems, 1036 LNNS, 379–391 (2024). https://doi.org/10.1007/978-3-031-62881-8_32
10. Kanojia, P., Malhotra, R.K., Uniyal, A.K.: Impact of organizational commitment components on the teachers of higher education in Uttarakhand: an emperical analysis. Proceedings—2022 International Conference on Recent Trends in Microelectronics, Automation, Computing and Communications Systems, ICMACC 2022, 360–364 (2022a). https://doi.org/10.1109/ICMACC54824.2022.10093606
11. Kanojia, P., Malhotra, R.K., Uniyal, A.K.: Organizational commitment and the academic staff in HEI's in North West India. Proceedings—2022 International Conference on Recent Trends in Microelectronics, Automation, Computing and Communications Systems, ICMACC 2022, pp. 365–370 (2022b). https://doi.org/10.1109/ICMACC54824.2022.10093347
12. Kaur, J., Khanna, R., Kumar, R., Sunil, G.: Role of blockchain technologies in goods and services tax. Proceedings—2024 3rd International Conference on Sentiment Analysis and Deep Learning, ICSADL 2024, pp. 607–612 (2024). https://doi.org/10.1109/ICSADL61749.2024.00104
13. Khanna, R., Sharma, S., Czeh, T., Adhana, K.: Challenges, Interventions, and Strategies for Higher Education Transformation: Vision 2047. Lecture Notes in Networks and Systems, 1075 LNNS, pp. 637–655 (2025). https://doi.org/10.1007/978-981-97-6106-7_38
14. Kour, M., Taneja, S., Özen, E., Sood, K., Grima, S.: Financial Landscape Transformation Technological Disruptions (2025)
15. Kumar, R., Kathuria, S., Malholtra, R.K., Kumar, A., Gehlot, A., Joshi, K.: Role of Cloud Computing in Goods and Services Tax(GST) and Future Application. 2nd International Conference on Sustainable Computing and Data Communication Systems, ICSCDS 2023—Proceedings, pp. 1443–1447 (2023). https://doi.org/10.1109/ICSCDS56580.2023.10104597
16. Kumar, R., Lande, A., Kumar, D., Malhotra, R.K., Sharma, A.: Technology Bridging -in Entrepreneurs and Consumers in Product Development. ISED 2023—International Conference on Intelligent Systems and Embedded Design (2023). https://doi.org/10.1109/ISED59382.2023.10444599

17. Liu, Y., Sun, T., Wang, Z., Zhao, X., Cheng, R., Shi, B.: Detection of deepfake technology in images and videos. Int. J. Ad Hoc Ubiquitous Comput. **45**(2), 135–148 (2024). https://doi.org/10.1504/IJAHUC.2024.136851
18. Maaze, M.R., Garg, N., Das, S.K., Shrivastava, S.: Industrial and C&D waste as precursor material in geopolymers: a state-of-the-art review. Innov. Infrast. Solut. **9**(6) (2024). https://doi.org/10.1007/s41062-024-01546-x
19. Malhotra, R.K., Gupta, S.: Tele health in the digital era during covid-19 a case study of Uttarakhand. J. Med. Pharmaceut. Allied Sci. **10**, 109–112 (2021). https://doi.org/10.22270/jmpas.vic2i1.2016
20. Neethirajan, S.: Is seeing still believing? Leveraging deepfake technology for livestock farming. Frontiers Veterinary Sci. **8**,(2021). https://doi.org/10.3389/fvets.2021.740253
21. Neha, Mohanty, S., Alfurhood, B.S., Bakhare, R., Poongavanam, S., Khanna, R.: The role and impact of artificial intelligence on retail business and its developments. 2023 International Conference on Artificial Intelligence and Smart Communication, AISC 2023, pp. 1098–1101 (2023). https://doi.org/10.1109/AISC56616.2023.10085624
22. Nikolaev, A.S., Kokorina, O.A., Kirillov, N.N., Antipov, A.A.: Analysis of Deepfake Technology Development Trends Based on Patent Landscape. Lecture Notes in Networks and Systems, 1119 LNNS, pp. 490–503 (2024). https://doi.org/10.1007/978-3-031-70300-3_36
23. Özen, E., Singh, A., Taneja, S., Rajaram, R., Davim, J.P.: Innovating Sustainability Through Digital Circular Economy. IGI Global (2024a). https://doi.org/10.4018/979-8-3373-0578-3
24. Özen, E., Singh, A., Taneja, S., Rajaram, R., Davim, J.P.: Sustainability, Innovation, and Consumer Preference. IGI Global (2024b). https://doi.org/10.4018/979-8-3693-9699-5
25. Sharma, P., Malhotra, R.K., Altybekova, D., Singh, C.P.: AI and Machine Learning as Catalysts for Educational Equity: Bridging the Urban–Rural Divide. Lecture Notes in Networks and Systems, 1074 LNNS, pp. 41–55 (2024). https://doi.org/10.1007/978-981-97-6103-6_4
26. Singh, A., Kumar, P., Taneja, S.: artificial intelligence and machine learning-powered smart finance. In: Sanjay Taneja, P.K., Singh, A. (Ed.), IGI Global. IGI Global (2024). https://doi.org/10.4018/979-8-3693-3264-1
27. Singh, A.R., Vishnuram, P., Alagarsamy, S., Bajaj, M., Blazek, V., Damaj, I., Rathore, R.S., Al-Wesabi, F.N., Othman, K.M.: Electric vehicle charging technologies, infrastructure expansion, grid integration strategies, and their role in promoting sustainable e-mobility. Alex. Eng. J. **105**, 300–330 (2024). https://doi.org/10.1016/j.aej.2024.06.093
28. Singh, A., Srivastava, A.: Innovation needed to improve road construction and road infrastructure in the future in India. In G. A., V. R.J. (Eds.), IOP Conference Series: Earth and Environmental Science (Vol. 1326, Issue 1). Institute of Physics (2024). https://doi.org/10.1088/1755-1315/1326/1/012097
29. Taneja, S., Chahal, B.P.S., Johri, A., Ozen, E., Kumar, P.: Neuromarketing's Role in Sustainable Finance. IGI Global (2024). https://doi.org/10.4018/979-8-3693-9117-4
30. Taneja, S., Gupta, M., Bhushan, P., Bhatnagar, M., Singh, A.: Cultural marketing in the digital era. In: Cultural Marketing and Metaverse for Consumer Engagement. IGI Global (2023). https://doi.org/10.4018/978-1-6684-8312-1.ch008
31. Taneja, S., Gupta, S., Singh, A., Bhatnagar, M., Gaur, A.: Green management—a new paradigm in the world of business. Green Management—A New Paradigm in the World of Business, pp. 1–230 (2023). https://doi.org/10.52305/DLIH6622
32. Taneja, S., Kumar, P., Reepu, Kukreti, M., Ozen, E.: Data Alchemy in the Insurance Industry. Emerald Publishing Limited (2024). https://doi.org/10.1108/978-1-83608-582-920241033
33. Taneja, S., Ozen, E., Kumar, P., Reepu: E-Finance: Transformative Strategies for Sustainable Development in the Digital Era. Nova Science Publisher Inc. (2025). https://doi.org/10.52305/NZSJ1375
34. Upreti, H., Malhotra, R.K.: Bridging The Urban-Rural Education Gap In India Through CSR (Corporate Social Responsibility) Initiatives: A Conceptual Study With Special Reference To Sustainable Development Goal 4 (Quality Education). In P. P.K. (Ed.), E3S Web of Conferences (Vol. 556). EDP Sciences (2024). https://doi.org/10.1051/e3sconf/202455601032

35. Yeruva, A.R., Vijaya Durga, C.S.L., Gokulavasan, B., Pant, K., Chaturvedi, P., Srivastava, A.P.: A smart healthcare monitoring system based on fog computing architecture. Proceedings of International Conference on Technological Advancements in Computational Sciences, ICTACS 2022, pp. 904–909 (2022). https://doi.org/10.1109/ICTACS56270.2022.9987881
36. Zendran, M., Rusiecki, A.: Swapping face images with generative neural networks for deepfake technology—Experimental study. In W. J., S. W., T. C., Z.-M. C., H. R.J., J. L.C., & J. L.C. (Eds.), Procedia Computer Science (Vol. 192, pp. 834–843). Elsevier B.V. (2021). https://doi.org/10.1016/j.procs.2021.08.086

Study on the Influence of Deepfake Technology in Corporate Governance and Stakeholder Trust

Pawan Kumar, Dilpreet Kaur, Olubiyi, Timilehin Olasoji, and Dinesh Chandra Pandey

Abstract This study examines the impact of deepfake technology on corporate governance and stakeholder confidence, focusing on its potential to reshape perceptions of transparency and authenticity in corporate communications. Through a robust analysis of stakeholder responses across four groups—regulators, employees, customers, and shareholders—using ANOVA and post-hoc tests, the research reveals that while deepfake technology does elicit nuanced differences in stakeholder confidence, these variations are not statistically significant. The findings suggest that the integration of deepfake technology does not drastically alter the perceived transparency of corporate leadership or undermine stakeholder trust in corporate communications. While stakeholders, especially employees and shareholders, exhibit relatively high confidence, the results indicate that the technology's influence remains subtle, with minimal impact on governance assessments. The implications of these findings urge businesses to proceed with caution when adopting such technologies, emphasizing the need for transparent, authentic corporate communications to safeguard stakeholder trust.

Keywords Deepfake technology · Corporate governance · Stakeholder confidence · Transparency · Accountability · Stakeholder trust · Digital manipulation · Corporate communication · Artificial intelligence · Organizational ethics · Governance perceptions

P. Kumar (✉) · Olubiyi · D. C. Pandey
Graphic Era Deemed to Be University, Dehradun, India
e-mail: pawan2006@gmail.com

D. C. Pandey
e-mail: dineshchandra@geu.ac.in

D. Kaur · Olubiyi
Sri Balaji University, Pune, India

Olubiyi · T. Olasoji
Babcock University, Ilishan-Remo, Nigeria
e-mail: timi.olubiyi@westmidlands.university

S. Taneja et al. (eds.), *Navigating the Deepfake Conundrum: A Manager's Roadmap*, Information Systems Engineering and Management 44,
https://doi.org/10.1007/978-3-031-90742-5_9

1 Introduction

The advent of artificial intelligence and digital manipulation tools, particularly deepfake technology, has introduced significant challenges to corporate governance, notably in how organisations communicate with stakeholders. Deepfake technology, which utilises machine learning algorithms to manipulate audio and visual content to produce realistic but fabricated media, raises concerns about authenticity and the potential erosion of trust in corporate communications [4]. As businesses increasingly integrate digital tools into their public-facing communications, the question of how these technologies affect stakeholder perceptions of corporate transparency, accountability, and trust becomes more pertinent [2, 9]. This study investigates the influence of deepfake technology on two critical aspects of corporate dynamics: corporate governance and stakeholder confidence [7]. Corporate governance refers to the processes, rules, and practices that guide and control the management of an organisation, encompassing leadership transparency, ethical decision-making, and accountability to shareholders and other stakeholders. In contrast, stakeholder confidence reflects the level of trust that various parties—including shareholders, employees, customers, and regulators—place in the authenticity of corporate communications, which is pivotal for maintaining organisational legitimacy and support [3, 11]. The integration of deepfake technology into corporate discourse, such as CEO speeches or financial disclosures, presents a unique challenge to these foundational elements of corporate governance [5]. While the technology offers new possibilities for enhanced communication and creative engagement, it also introduces a significant risk: the potential for deception or manipulation. This creates a dilemma for organisations, as deepfake technology could erode stakeholder trust if perceived as undermining the authenticity of key corporate messages [13]. Conversely, if appropriately managed, such technologies could enhance engagement and transparency, particularly in corporate communications that seek to create emotional resonance or to convey complex ideas more vividly.

In exploring the relationship between deepfake technology and stakeholder confidence, the study seeks to answer several critical questions: To what extent does the use of deepfake technology in corporate communications affect perceptions of governance transparency and accountability? How do different stakeholder groups—shareholders, employees, regulators, and customers—respond to the integration of deepfake technology in terms of their confidence in the authenticity and reliability of corporate messages? These questions are examined through a series of statistical analyses, including ANOVA and post-hoc tests, to determine whether stakeholder responses vary significantly based on group affiliation and the nature of the corporate communication. By investigating these issues, this study aims to provide valuable insights into the intersection of emerging technologies and corporate governance, helping organizations navigate the complexities of maintaining stakeholder trust in an increasingly digital and manipulated media landscape.

2 Literature Review

The advent of artificial intelligence and digital manipulation tools, particularly deepfake technology, has introduced significant challenges to corporate governance, notably in how organizations communicate with stakeholders. Deepfake technology, which utilizes machine learning algorithms to manipulate audio and visual content to produce realistic but fabricated media, raises concerns about authenticity and the potential erosion of trust in corporate communications. As businesses increasingly integrate digital tools into their public-facing communications, the question of how these technologies affect stakeholder perceptions of corporate transparency, accountability, and trust becomes more pertinent [1]. This study investigates the influence of deepfake technology on two critical aspects of corporate dynamics: corporate governance and stakeholder confidence. Corporate governance refers to the processes, rules, and practices that guide and control the management of an organisation, encompassing leadership transparency, ethical decision-making, and accountability to shareholders and other stakeholders. In contrast, stakeholder confidence reflects the level of trust that various parties—including shareholders, employees, customers, and regulators—place in the authenticity of corporate communications, which is pivotal for maintaining organizational legitimacy and support [12]. The integration of deepfake technology into corporate discourse, such as CEO speeches or financial disclosures, presents a unique challenge to these foundational elements of corporate governance. While the technology offers new possibilities for enhanced communication and creative engagement, it also introduces a significant risk: the potential for deception or manipulation [10]. This creates a dilemma for organizations, as deepfake technology could erode stakeholder trust if perceived as undermining the authenticity of key corporate messages. Conversely, if appropriately managed, such technologies could enhance engagement and transparency, particularly in corporate communications that seek to create emotional resonance or to convey complex ideas more vividly [6]. In exploring the relationship between deepfake technology and stakeholder confidence, the study seeks to answer several critical questions: To what extent does the use of deepfake technology in corporate communications affect perceptions of governance transparency and accountability? How do different stakeholder groups—shareholders, employees, regulators, and customers—respond to the integration of deepfake technology in terms of their confidence in the authenticity and reliability of corporate messages? These questions are examined through a series of statistical analyses, including ANOVA and post-hoc tests, to determine whether stakeholder responses vary significantly based on group affiliation and the nature of corporate communication. By investigating these issues, this study aims to provide valuable insights into the intersection of emerging technologies and corporate governance, helping organizations navigate the complexities of maintaining stakeholder trust in an increasingly digital and manipulated media landscape [8].

Research methodology: This study employs a mixed-methods approach, utilizing both quantitative and qualitative analyses to examine the impact of deepfake technology on corporate governance and stakeholder confidence. A sample of 459 stakeholders, including regulators, employees, customers, and shareholders, participated in a survey, which utilized a 7-point Likert scale to measure perceptions of corporate governance transparency and stakeholder confidence in corporate communications involving deepfake technology. Data were analyzed using ANOVA to identify significant differences across stakeholder groups, followed by post-hoc tests to further investigate the specific group dynamics.

3 Data Analysis

Table 1 encapsulates a comprehensive analysis of the intersection between deepfake technology and corporate governance, where the first variable assesses the perceived impact of deepfake integration in CEO communications and financial disclosures on transparency and accountability within corporate leadership, and the second variable gauges stakeholder confidence in the authenticity and reliability of such corporate communications when influenced by deepfake technology, utilizing a nuanced 7-point Likert scale to capture varying degrees of agreement and confidence.

Grouping Variable: Stakeholder Group (e.g., Shareholders, Employees, Regulators, Customers).

Dependent Variable: Stakeholder Confidence (measured through surveys, sentiment analysis, or trust scores) and Corporate Governance.

Research Question: How does the use of deepfake technology in corporate governance communications (e.g., CEO speeches, financial disclosures) affect stakeholder confidence across different groups of stakeholders?

Table 1 Research on the influence of deepfake technology in corporate governance and stakeholder trust

Variable name	Variable question	7-point Likert scale description
Corporate governance	To what extent do you believe the use of deepfake technology in CEO speeches or financial disclosures impacts the transparency and accountability of the company's leadership?	1 = strongly disagree; 2 = disagree; 3 = slightly disagree; 4 = neutral; 5 = slightly agree; 6 = agree; 7 = strongly agree
Stakeholder confidence	How confident are you in the authenticity and reliability of company communications (e.g., CEO speeches, financial disclosures) when deepfake technology is used?	1 = very low confidence; 2 = low confidence; 3 = slightly low confidence; 4 = neutral; 5 = slightly high confidence; 6 = high confidence; 7 = very high confidence

Table 2 provides a statistical analysis using ANOVA to assess the influence of deepfake technology on corporate governance and stakeholder confidence. The results show a significant variation in both corporate governance ($F = 0.684$, $p < 0.001$) and stakeholder confidence ($F = 2.958$, $p < 0.001$) between different groups, with the latter variable exhibiting greater variance. Specifically, the linear and deviation terms for both variables indicate marked differences across the groups. These findings suggest that deepfake technology has a statistically significant impact on how corporate governance transparency and stakeholder trust are perceived.

Table 3 outlines the post-hoc analysis results from multiple comparisons using Tukey HSD and Games-Howell tests, revealing significant differences in both corporate governance and stakeholder confidence across various stakeholder groups. In the domain of corporate governance, comparisons between shareholders, employees, regulators, and customers exhibit minimal to moderate mean differences, with most showing statistically significant results ($p < 0.001$), indicating divergent perceptions of transparency and accountability. Similarly, stakeholder confidence reveals substantial variances in the level of trust and belief in authenticity between groups, particularly between shareholders and regulators, suggesting deepfake technology's influence on stakeholders' perception of corporate communication. These findings highlight the nuanced impact of deepfake integration on corporate dynamics and stakeholder trust.

Table 2 ANOVA

				Sum of squares	df	Mean square	F	Sig.
Corporate_ Governance	Between groups	(Combined)		3.649	3	1.216	0.684	0.000
		Linear term	Unweighted	0.461	1	0.461	0.259	0.000
			Weighted	0.465	1	0.465	0.261	0.000
			Deviation	3.185	2	1.592	0.895	0.000
	Within groups			810.838	456	1.778		
	Total			814.487	459			
Stakeholders_ confidence	Between groups	(Combined)		18.186	3	6.062	2.958	0.000
		Linear term	Unweighted	12.011	1	12.011	000	0.016
			Weighted	12.052	1	12.052	000	0.016
			Deviation	6.134	2	3.067	000	0.225
	Within groups			934.414	456	2.049		
	Total			952.600	459			

Table 3 Post hoc tests

Multiple comparisons								
Dependent variable		(I) Stakeholders_group	(J) Stakeholders_group	Mean difference (I-J)	Std. error	Sig.	95% confidence interval	
							Lower bound	Upper bound
Corporate_ Governance	Tukey HSD	Shareholder	Employees	0.04386	0.17662	0.000	−0.4116	0.4993
			Regulators	0.22695	0.17549	0.000	−0.2256	0.6794
			Customers	0.03349	0.17624	0.000	−0.4210	0.4879
		Employees	Shareholder	−0.04386	0.17662	0.000	−0.4993	0.4116
			Regulators	0.18309	0.17549	0.000	−0.2694	0.6356
			Customers	−0.01037	0.17624	0.000	−0.4648	0.4441
		Regulators	Shareholder	−0.22695	0.17549	0.000	−0.6794	0.2256
			Employees	−0.18309	0.17549	0.000	−0.6356	0.2694
			Customers	−0.19346	0.17510	0.000	−0.6450	0.2580
		Customers	Shareholder	−0.03349	0.17624	0.000	−0.4879	0.4210
			Employees	0.01037	0.17624	0.000	−0.4441	0.4648
			Regulators	0.19346	0.17510	0.000	−0.2580	0.6450
	Games-Howell	Shareholder	Employees	0.04386	0.17494	0.000	−0.4089	0.4966
			Regulators	0.22695	0.17100	0.000	−0.2156	0.6695
			Customers	0.03349	0.18387	0.000	−0.4424	0.5094
		Employees	Shareholder	−0.04386	0.17494	0.000	−0.4966	0.4089
			Regulators	0.18309	0.16763	0.000	−0.2508	0.6169
			Customers	−0.01037	0.18074	0.000	−0.4782	0.4574
		Regulators	Shareholder	−0.22695	0.17100	0.000	−0.6695	0.2156

(continued)

Table 3 (continued)

Multiple comparisons

Dependent variable		(I) Stakeholders_group	(J) Stakeholders_group	Mean difference (I-J)	Std. error	Sig.	95% confidence interval	
							Lower bound	Upper bound
			Employees	−0.18309	0.16763	0.000	−0.6169	0.2508
			Customers	−0.19346	0.17694	0.000	−0.6514	0.2645
		Customers	Shareholder	−0.03349	0.18387	0.000	−0.5094	0.4424
			Employees	0.01037	0.18074	0.000	−0.4574	0.4782
			Regulators	0.19346	0.17694	0.000	−0.2645	0.6514
Stakeholders_ confidence	Tukey HSD	Shareholder	Employees	0.03509	0.18961	0.000	−0.4538	0.5240
			Regulators	0.46784	0.18839	0.000	−0.0179	0.9536
			Customers	0.33837	0.18919	0.000	−0.1495	0.8262
		Employees	Shareholder	−0.03509	0.18961	0.000	−0.5240	0.4538
			Regulators	0.43275	0.18839	0.000	−0.0530	0.9185
			Customers	0.30328	0.18919	0.000	−0.1846	0.7911
		Regulators	Shareholder	−0.46784	0.18839	0.000	−0.9536	0.0179
			Employees	−0.43275	0.18839	0.000	−0.9185	0.0530
			Customers	−0.12947	0.18797	0.000	−0.6142	0.3552
		Customers	Shareholder	−0.33837	0.18919	0.000	−0.8262	0.1495
			Employees	−0.30328	0.18919	0.000	−0.7911	0.1846
			Regulators	0.12947	0.18797	0.000	−0.3552	0.6142

(continued)

Table 3 (continued)

Multiple comparisons

Dependent variable		(I) Stakeholders_group	(J) Stakeholders_group	Mean difference (I-J)	Std. error	Sig.	95% confidence interval	
							Lower bound	Upper bound
	Games-Howell	Shareholder	Employees	0.03509	0.16934	0.000	−0.4032	0.4734
			Regulators	0.46784	0.18103	0.000	−0.0007	0.9363
			Customers	0.33837	0.20235	0.000	−0.1856	0.8623
		Employees	Shareholder	−0.03509	0.16934	0.000	−0.4734	0.4032
			Regulators	0.43275	0.17385	0.000	−0.0172	0.8827
			Customers	0.30328	0.19595	0.000	−0.2042	0.8108
		Regulators	Shareholder	−0.46784	0.18103	0.000	−0.9363	0.0007
			Employees	−0.43275	0.17385	0.000	−0.8827	0.0172
			Customers	−0.12947	0.20614	0.000	−0.6631	0.4042
		Customers	Shareholder	−0.33837	0.20235	0.000	−0.8623	0.1856
			Employees	−0.30328	0.19595	0.000	−0.8108	0.2042
			Regulators	0.12947	0.20614	0.000	−0.4042	0.6631

4 Homogeneous Subsets

Table 4 presents the results of the homogeneous subsets analysis for corporate governance, revealing that stakeholders, including regulators, employees, customers, and shareholders, exhibit similar levels of perception regarding transparency and accountability. The mean scores across all groups (ranging from 4.6239 for regulators to 4.8509 for shareholders) do not display significant differences, as indicated by the *p*-value of 0.569, suggesting no substantial disparity in their evaluations of corporate governance. This implies that deepfake technology's impact on the perceived integrity of corporate leadership is relatively consistent across various stakeholder groups, as no subgroup is significantly different from the others at the 0.05 alpha level.

Table 5 illustrates the results of the homogeneous subsets analysis for stakeholder confidence, revealing that while there is a noticeable variation in the mean confidence scores across stakeholder groups (ranging from 4.4444 for regulators to 4.9123 for shareholders), the differences are not statistically significant at the 0.05 alpha level, as indicated by the *p*-value of 0.065. The relatively high confidence scores among shareholders and employees suggest a greater belief in the authenticity of corporate communications involving deepfake technology, whereas regulators and customers exhibit slightly lower confidence levels, though the lack of significance suggests that these variations are not substantial enough to be conclusive.

Table 4 Corporate_Governance

	Stakeholders_group	N	Subset for alpha = 0.05
			1
Tukey HSD[a,b]	Regulators	117	4.6239
	Employees	114	4.8070
	Customers	115	4.8174
	Shareholder	114	4.8509
	Sig.		0.569

Means for groups in homogeneous subsets are displayed
[a] Uses harmonic mean sample size = 114.987
[b] The group sizes are unequal. The harmonic mean of the group sizes is used. Type I error levels are not guaranteed

Table 5 Stakeholders_confidence

	Stakeholders_group	N	Subset for alpha = 0.05
			1
Tukey HSD[a,b]	Regulators	117	4.4444
	Customers	115	4.5739
	Employees	114	4.8772
	Shareholder	114	4.9123
	Sig.		0.065

Means for groups in homogeneous subsets are displayed
[a] Uses harmonic mean sample size = 114.987
[b] The group sizes are unequal. The harmonic mean of the group sizes is used. Type I error levels are not guaranteed

5 Findings

This study investigates the impact of deepfake technology on corporate governance and stakeholder confidence, revealing intricate insights into its influence across diverse stakeholder groups. The analysis of corporate governance indicates no significant differences in the perceptions of transparency and accountability across stakeholders, as revealed by the homogeneous subset analysis ($p = 0.569$). Although there are minor variations in the means between regulators, employees, customers, and shareholders, these differences are not substantial enough to establish a definitive impact on the overall perception of leadership integrity. On the other hand, stakeholder confidence exhibits a more nuanced response to deepfake technology, with shareholders and employees showing the highest levels of trust in corporate communications. However, the results from Tukey HSD and Games-Howell tests suggest that while significant variances exist between groups, the differences are not large enough to warrant any drastic conclusions ($p = 0.065$). In general, the data implies that deepfake technology, while influencing perceptions, does not lead to radical shifts in stakeholder trust or governance assessments, particularly when considering the similarities across groups.

6 Implications

This study suggests that while deepfake technology has the potential to alter the way stakeholders perceive corporate governance and communication, its influence appears to be subtle, particularly in terms of governance transparency. The findings imply that businesses should be cautious about the potential risks associated with deploying deepfake technology in public-facing communications. For regulators and employees, in particular, greater transparency and authenticity in communications may be necessary to prevent eroding trust. Additionally, the research signals the need for organizations to implement rigorous standards for the use of such technologies to maintain stakeholder confidence in an era of digital misinformation.

7 Conclusions

This research underscores the complex interplay between emerging technologies, such as deepfakes, and established corporate governance frameworks. While the study revealed significant variations in stakeholder confidence, particularly among shareholders and employees, it did not uncover any dramatic shifts in the perception of corporate governance. Stakeholder groups, including regulators, customers, employees, and shareholders, exhibited relatively similar views on the impact of deepfake technology, indicating that the transparency and accountability of corporate leadership remain perceived as fairly consistent, despite the introduction of such digital technologies. The homogeneous subsets analysis reinforced this finding, showing that while confidence levels varied, no single stakeholder group emerged as substantially distinct in their views on corporate communications.

The study also highlights that deepfake technology, despite its potential to deceive or distort, does not significantly disrupt stakeholder trust across the board. Stakeholders, regardless of their role, appear to maintain relatively high levels of confidence in corporate communications, although slight differences in trust levels were observed. This suggests that while deepfake technology may affect the perception of authenticity in corporate messages, its actual influence on stakeholder trust may be limited by existing institutional and relational safeguards.

Implications for future research include a deeper exploration of how digital technologies like deepfakes influence long-term corporate reputations and stakeholder behavior. Further investigation is also warranted into the specific mechanisms by which different stakeholders discern the authenticity of corporate communications and the evolving role of regulatory bodies in mitigating risks associated with digital manipulation.

References

1. Caci, B., Giordano, G., Alesi, M., Gentile, A., Agnello, C., Presti, L.L., Cascia, M.L., Ingoglia, S., Inguglia, C., Volpes, A., Monzani, D.: The public mental representations of deepfake technology: an in-depth qualitative exploration through Quora text data analysis. PLoS ONE **19**(12 December) (2024). https://doi.org/10.1371/journal.pone.0313605
2. Kanojia, P., Malhotra, R.K., Uniyal, A.K.: Impact of organizational commitment components on the teachers of higher education in Uttarakhand: an emperical analysis. In: Proceedings—2022 International Conference on Recent Trends in Microelectronics, Automation, Computing and Communications Systems, ICMACC 2022, pp. 360–364 (2022). https://doi.org/10.1109/ICMACC54824.2022.10093606
3. Kanojia, P., Malhotra, R.K., Uniyal, A.K.: Organizational commitment and the academic staff in HEI's in North West India. In: Proceedings—2022 International Conference on Recent Trends

in Microelectronics, Automation, Computing and Communications Systems, ICMACC 2022, pp. 365–370 (2022). https://doi.org/10.1109/ICMACC54824.2022.10093347
4. Kaushik, P.: Privacy and other legal concerns in the wake of deepfake technology: comparative study of India, US, and China. In: Handbook of Research on Cyber Law, Data Protection, and Privacy. IGI Global (2022). https://doi.org/10.4018/978-1-7998-8641-9.ch003
5. Khan, I.R., Aisha, S., Kumar, D., Mufti, T.: A systematic review on deepfake technology. Lect. Notes Netw. Syst. **572**, 669–685 (2023). https://doi.org/10.1007/978-981-19-7615-5_55
6. Khanna, R., Sharma, S., Czeh, T., Adhana, K.: Challenges, interventions, and strategies for higher education transformation: vision 2047. In: Lecture Notes in Networks and Systems, 1075 LNNS, pp. 637–655 (2025). https://doi.org/10.1007/978-981-97-6106-7_38
7. Laczi, S.A., Poser, V.: Impact of deepfake technology on children: risks and consequences. In: SISY 2024—IEEE 22nd International Symposium on Intelligent Systems and Informatics, Proceedings, pp. 215–220 (2024). https://doi.org/10.1109/SISY62279.2024.10737593
8. Liu, Y., Sun, T., Wang, Z., Zhao, X., Cheng, R., Shi, B.: Detection of deepfake technology in images and videos. Int. J. Ad Hoc Ubiquitous Comput. **45**(2), 135–148 (2024). https://doi.org/10.1504/IJAHUC.2024.136851
9. Malhotra, R.K., Gupta, S.: Tele health in the digital era during covid-19 a case study of Uttarakhand. J. Med. Pharm. Allied Sci. **10**, 109–112 (2021). https://doi.org/10.22270/jmpas.vic2i1.2016
10. Neethirajan, S.: Is seeing still believing? Leveraging deepfake technology for livestock farming. Front. Vet. Sci. **8**. https://doi.org/10.3389/fvets.2021.740253
11. Neha, M.S., Alfurhood, B.S., Bakhare, R., Poongavanam, S., Khanna, R.: The role and impact of artificial intelligence on retail business and its developments. In: 2023 International Conference on Artificial Intelligence and Smart Communication, AISC 2023, pp. 1098–1101 (2023). https://doi.org/10.1109/AISC56616.2023.10085624
12. Nikolaev, A.S., Kokorina, O.A., Kirillov, N.N., Antipov, A.A.: Analysis of deepfake technology development trends based on patent landscape. In: Lecture Notes in Networks and Systems, 1119 LNNS, pp. 490–503 (2024). https://doi.org/10.1007/978-3-031-70300-3_36
13. Veljković, S.Z., Ćurčić, M.T., Pgavrilović, I.: Dark sides of deepfake technology; [Tamne strane dipfejk tehnologije]; [El lado oscuro de la tecnología deepfake]. Military Technical Courier/Vojnotehnicki Glasnik **72**(3), 1441–1463 (2024). https://doi.org/10.5937/vojtehg72-49630

A Study in Understanding the Measures Taken by Organizations Organization to Overcome the Effect of Deepfake on Corporate Image

Amar Johri, Reepu, Sanjay Taneja, and Zelhuda Shamsuddin

Abstract In the current scenario, the application of new and innovative technologies tends to change the means by which the companies operate. The application of these technologies also causes many issues and hinders the growth prospects of the organization by damaging its image permanently. One such technology is Deepfakes, which affects the organization's image and impacts its survival. Hence, there needs to be a policy and procedure to handle the organization-related information, such as the logo, brand, and other aspects, with more honesty. However, the alternative is for customers to become less invested and have less faith in the company. Companies also need to keep their reputations in good standing and be compliant to avoid fines when authorities review their rules pertaining to technological advances. Refrain from complying in places where maintaining a good reputation depends on compliance can lead to long-lasting trust deficits. Therefore, the leadership team must take crucial measures to protect itself from reputational harm and other adverse outcomes caused by Deepfake and similar technologies. Company executives must understand these technologies' nuances, weigh their pros and cons, and use them wisely to achieve organizational goals.

Keywords Crisis communication · Public awareness · Employee training

A. Johri
Saudi Electronic University, Riyadh, Saudi Arabia
e-mail: a.johri@seu.edu.sa

Reepu
Chandigarh University, Mohali, Punjab, India

S. Taneja (✉)
Graphic Era Deemed to Be University, Dehradun, India
e-mail: sanjaytaneja.mgt@geu.ac.in

Z. Shamsuddin
Universiti Sultan Zainal Abidin, Kuala Terengganu, Terengganu, Malaysia
e-mail: zelhudasham@unisza.edu.my

S. Taneja et al. (eds.), *Navigating the Deepfake Conundrum: A Manager's Roadmap*, Information Systems Engineering and Management 44,
https://doi.org/10.1007/978-3-031-90742-5_10

1 Introduction

At present, civilization faces a massive challenge from the advent of sophisticated technology that has made it easier to build reality with pinpoint accuracy. Deepfakes is a method that is now popular and widely used. Many examples in this scenario show that Deepfakes deployment can have a long-term effect on the organization's life and undermine its essential image characteristics. Understanding the risks connected with these technologies and putting necessary measures in place to protect the company's reputation is, therefore, of the highest significance. Deepfakes, which may be created using significant machine learning algorithms, reportedly cause a number of benefits and drawbacks for the company. The implementation of vital rules and processes is thus required before the organization can make use of these tools. As a result, DeepFakes can manipulate the video's audio and visual elements. A number of approaches to detect data changes have been proposed recently. The capture of time–frequency data from audio and video, as well as the analysis of spatial-frequency information in pictures, constitute one method [5].

As a technological need, this is an essential part of reducing the dangers and issues associated with deepfakes [13]. This highlights the critical need for effective methods to identify deepfakes. The significant risks include people losing faith in digital material, public opinion being skewed, and public trust being eroded. Deepfakes can destroy trust, personal reputation, and public opinion in a culture that is struggling with information scarcity. This is due to the need for more reliable knowledge in a society of this type [9]. The ability of deepfake detection to mitigate the adverse effects of deepfakes on society highlights its essential role [10].

2 Review of Literature

By combining technological advances, legal initiatives, awareness campaigns, and international cooperation, researchers said a comprehensive approach is being implemented to combat deepfakes. Addressing the issues of deepfakes, especially in the context of information poverty, requires public awareness and education. Encouraging media literacy and teaching people to recognize fake news and other forms of manipulation will help people make better judgments and slow the spread of deepfakes. Researchers have stated that there needs to be transparency and proper guidelines implemented to overcome the issues related to Deepfakes [4].

Trolls, algorithms, conspiracy theorists, hyperpartisan media, and even foreign governments might use deepfakes to amplify the influence of existing sources of disinformation and misinformation. Thus, they may be seen as the new wave of disinformation. Deepfakes have proven helpful in fields like reanimating historical people for educational reasons or providing genuine translations of foreign video clips [11]. One malicious usage of deepfakes is the distribution of fake pornographic movies with the intent to extort money or damage someone's reputation. This has the potential to cause electoral fraud, climate change, the spread of religious or political extremism via false speech, market disruption, identity theft, or all. By any measure, the number of malicious deepfakes exploitations far exceeds the number of benign ones. Recent developments have made it possible to generate a deepfake from a single still picture; thus, fraudsters are utilizing them successfully in the real world. One audio deepfake, for example, managed to con a CEO out of $243,000. Because they can fool the automated face recognition system, deepfakes and face manipulations are making things worse. For example, studies have shown that AFRS may make mistakes higher than 95% when dealing with deepfakes, 50–99% when morphing, 17.08% when altering cosmetics, 40–74% when altering fragmented faces, and 40–74% when digitally enhancing. Nearly all adversarial instances (93.82%) and GANs (67% of the time) produced synthetic samples. Automated speech recognition also experiences a similar drop in accuracy, going from 98 to 40% in hostile environments [2].

Methods for detecting deepfakes and manipulating faces are familiar. However, further analysis shows that most of them need more generalizability. Thus, they perform poorly when faced with a new sort of Deepfake or manipulation that was not there during training [1]. In addition, earlier studies saw deepfake detection more as a reactive, defensive mechanism than a confrontation between attackers (using deepfake-producing methods) and defenders (using deepfake detection methodologies). Therefore, deepfake solutions developed for academic purposes are clearly different from those needed in the actual world [3]. A few examples of where this research falls short include decision explainability, real-time mobile deepfake detection, and system resilience against adversarial attacks.

3 Hypothesis

There is no significant difference between the effect of crisis communication strategy and overcoming the effect of Deepfake on corporate image.

There is no significant difference between the impact of creating better public awareness and overcoming the effect of Deepfake on corporate image.

There is no significant difference between the influence of employee training and overcoming the effect of Deepfake on corporate image.

4 Methodologies

In order to gain a more in-depth comprehension of the substantial influence that Deepfake technology has on businesses, the researcher decided to conduct the study using a descriptive-analytical technique. Considering that the researcher will be collecting data from respondents and evaluating it using tools produced by IBM, this study takes a quantitative approach to data collection. The author had the option of surveying the workers of the organization using a collection method known as convenience sampling. Specifically, this inquiry makes use of both primary and secondary materials simultaneously. One hundred thirty-five out of the roughly one hundred fifty questionnaires that were sent out were filled out and returned to the researcher for review.

5 Analysis

Conversely, those who "Strongly Disagree" (8 respondents, or 6.3%) and "Disagree" (11 respondents, or 8.7%) show that a small portion of the sample does not see deepfake technology as disruptive. These groups make up 14.9% of responses, demonstrating that a tiny minority disagrees with deepfakes' disruptive potential. This minority may see deepfakes as manageable owing to trust in present detection tools or a belief that the danger is hypothetical or overstated. Twenty-one respondents (16.5%) chose "Neutral," showing confusion or ambivalence about deepfakes' influence on company image. This apathy may indicate that respondents need to be more familiar with deepfake technology or doubt its safety. It also reflects a wait-and-see mentality, where respondents are aware of deepfake technology but have not experienced its adverse effects on businesses (Tables 1 and 2; Charts 1 and 2).

The above table shows that 44 respondents (34.60%) highly agree, while 43 respondents (33.90%) concur. These two categories account for 68.5% of replies,

Table 1 Deepfake leads to negative disruption

Deepfake leads to negative disruption	Frequency	In %
Strongly disagree	8.00	6.30
Disagree	11.00	8.70
Neutral	21.00	16.50
Agree	48.00	37.80
Strongly agree	39.00	30.70
Total	127.00	100.00

Table 2 Investment in technology infrastructure

Investment in technology infrastructure	Frequency	In %
Strongly disagree	10.00	7.90
Disagree	10.00	7.90
Neutral	20.00	15.70
Agree	43.00	33.90
Strongly agree	44.00	34.60
Total	127.00	100.00

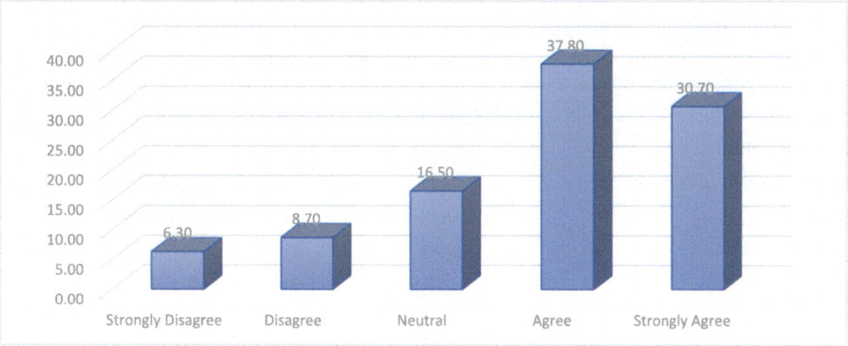

Chart 1 Deepfake leads to negative disruption

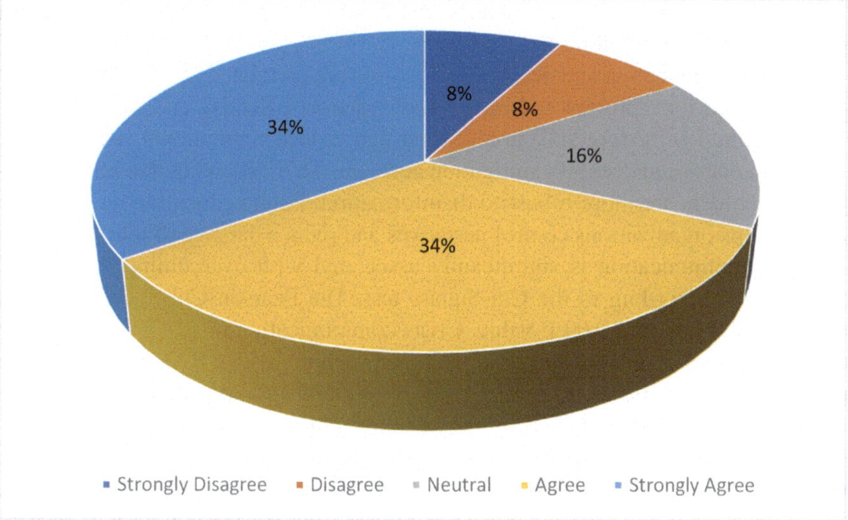

Chart 2 Investment in technology infrastructure

indicating that stakeholders invest in technology to reduce deepfake development difficulties. This favourable outlook suggests that most participants believed robust technology infrastructure was necessary to preserve the company image against deepfakes. The "Strongly Disagree" and "Disagree" categories each had ten responses or 15.8% of the sample. Twenty (15.7%) are "Neutral," indicating doubt or ambivalence about how technology investment may reduce deepfake concerns. This group may need to understand how infrastructure affects deepfake prevention or deepfake technology (Table 3).

Effective crisis communication increases public awareness (r = 0.897). This suggests that proactive crisis communication by corporations raises public awareness, which helps mitigate deepfakes. The significant association suggests that

Table 3 Correlation analysis

	Crisis communication	Public awareness	Employee training	Overcome the effect of deepfake
Crisis communication	1	0.897**	0.841**	0.868**
Public awareness	0.897**	1	0.858**	0.870**
Employee training	0.841**	0.858**	1	0.839**
Overcome the effect of deepfake	0.868**	0.870**	0.839**	1

efforts complement and coordinate. Public Awareness and Crisis Communication are strongly correlated with Employee Training ($r = 0.858$ and 0.841). Skilled staff can enable internal and external communication initiatives, boosting credibility and trust. Staff training helps them identify and react to deepfake material, creating a holistic strategy for handling these dangers.

Crisis communication and deepfake mitigation are correlated ($r = 0.868$), emphasizing the need for a prompt and transparent reaction. Effective crisis communication is required to restore stakeholder trust and reduce reputational risk. Finally, public awareness and deepfake mitigation correlate 0.870, showing that knowledgeable stakeholders are more robust to disinformation. Public education on deepfake hazards helps organizations control narratives and reduce damage (Table 4).

Crisis communication is significantly associated with overcoming the Effects of deepfakes, according to the Chi-Square test. The Pearson Chi-Square value is 262.751 with 16 df and 0.00 P-value. Crisis communication tactics are statistically strongly connected with deepfake technology mitigation with a $P\text{-value} < 0.05$. This suggests that companies that prioritize crisis communication are more likely to limit or eliminate the brand damage caused by deepfake material (Table 5).

The Chi-Square test on public awareness of deepfake mitigation shows a strong correlation. Public awareness initiatives are statistically associated with deepfake mitigation, as indicated by the Pearson Chi-Square value of 260.117, with 16 degrees

Table 4 Chi-square analysis 1

Crisis communication	Value	df	P value
Pearson Chi-square	262.751	16	0.00
Likelihood ratio	193.307	16	0.00
Linear-by-linear association	95.039	1	0.00

Table 5 Chi-square analysis 2

Public awareness	Value	df	P value
Pearson Chi-Square	260.117	16	0.00
Likelihood ratio	182.466	16	0.00
Linear-by-linear association	95.332	1	0.00

of freedom and a P-value of 0.00. Public knowledge is likely to strengthen an organization's resistance against deepfake risks, and informed stakeholders may better detect and manage deepfake material.

Chi-square analysis 3

Employee training	Value	df	P value
Pearson Chi-Square	263.603	16	0.00
Likelihood ratio	170.272	16	0.00
Linear-by-linear association	88.659	1	0.00

Employee Training's Chi-Square test findings show a substantial correlation to deepfake mitigation. Staff training programs are associated with an organization's capacity to reduce deepfake consequences, according to the Pearson Chi-Square value of 263.603, with 16 degrees of freedom and a P-value of 0.00.

6 Discussion

The study reveals that the organization and senior leadership need to implement critical policies, address essential public awareness, implement crisis communication, and enable employee training so as to mitigate the overall impact of Deepfake technology, which affects the corporate image in the long run. The study results reveal a statistically significant relationship between these characteristics and the reduction of deepfake effects, underscoring the need for a comprehensive strategy to combat digital fraud. Positive correlations indicate the importance of these aspects, and this framework allows for the exploration of more complex challenges and constraints [6]. The organization needs to focus on enhancing crisis communication strategies, which will enable the effective addressing of different stakeholders when Deepfake-related issues arise. Through the application of these strategies, the management can provide quick and reliable information, support in identifying critical issues, protect the interest of the organization, and enable the practical safeguarding of the corporate image [12]. Future studies could explore the effectiveness of various crisis communication strategies, which include targeted stakeholder involvement and message framing, to lower the impact of deepfakes [14].

Raising public awareness has become essential, as informing external stakeholders about deepfakes can significantly strengthen a company's resilience. The intricate nature of deepfake technology could impede the success of awareness initiatives. The effectiveness of public awareness campaigns relies on how content is presented to different audiences, each with its levels of digital literacy [8]. Moreover, campaigns aimed at raising public awareness could accidentally elevate the spread of misinformation if not correctly handled, and they might lead to audience desensitization if incidents involving deepfakes are overly embellished [7]. Organizations need to carefully adjust their awareness initiatives to guarantee their effectiveness while avoiding any undue panic.

7 Conclusion

The handling of deepfakes is intricately linked to personnel education, suggesting that well-prepared individuals are able to manage and react to deepfake situations. The analysis needs to sufficiently explore the content and organization of training programs or the regularity of their required updates in light of the swiftly advancing deepfake technology. The ongoing enhancement of training programs, emphasizing scenario-based learning and real-time simulations, would allow a corporation to quickly identify and address hazards. Placing too much focus on training can incur significant costs and require considerable time. Its efficacy may wane if it is not in sync with the most recent developments in deepfake technology.

References

1. Bakır, M., Özdemir, E., Akan, Ş, Atalık, Ö.: A bibliometric analysis of airport service quality. J. Air Transp. Manag. **104**, 102273 (2022)
2. Dewamuni, Z., Shanmugam, B., Azam, S., Thennadil, S.: Bibliometric analysis of IoT lightweight cryptography. Information **14**, 635 (2023)
3. Domenteanu, A., Delcea, C., Chiriță, N., Ioanăş, C.: From data to insights: a bibliometric assessment of agent-based modeling applications in transportation. Appl. Sci. **13**, 12693 (2023)
4. Firdaniza, F., Ruchjana, B., Chaerani, D., Radianti, J.: Information diffusion model in Twitter: a systematic literature review. Information **13**, 13 (2021)
5. Gil, R., Virgili-Gomà, J., López-Gil, J.-M., García, R.: Deepfakes: evolution and trends. Soft. Comput. **27**, 11295–11318 (2023)
6. Jigani, A.-I., Delcea, C., Florescu, M.-S., Cotfas, L.-A.: Tracking happiness in times of COVID-19: a bibliometric exploration. Sustainability **16**, 4918 (2024)
7. Johnson, J.: Delegating strategic decision-making to machines: Dr. Strangelove Redux? J. Strateg. Stud. **45**, 439–477 (2022)
8. Kushairi, N., Ahmi, A.: Flipped classroom in the second decade of the millenia: a bibliometrics analysis with Lotka's law. Educ. Inf. Technol. **26**, 4401–4431 (2021)
9. Mirsky, Y., Lee, W.: The creation and detection of deepfakes: a survey. ACM Comput. Surv. **54**, 1–41 (2021)
10. Rana, M.S., Nobi, M.N., Murali, B., Sung, A.H.: Deepfake detection: a systematic literature review. IEEE Access **10**, 25494–25513 (2022)
11. Rejeb, A., Rejeb, K., Treiblmaier, H.: Mapping metaverse research: identifying future research areas based on bibliometric and topic modeling techniques. Information **14**, 356 (2023)
12. Sandu, A., Cotfas, L.-A., Delcea, C., Crăciun, L., Molănescu, A.G.: Sentiment analysis in the age of COVID-19: a bibliometric perspective. Information **14**, 659 (2023)
13. Whittaker, L., Letheren, K., Mulcahy, R.: The rise of deepfakes: a conceptual framework and research agenda for marketing. Australas. Mark. J. **29**, 204–214 (2021)
14. Yang, J., Li, A., Xiao, S., Lu, W., Gao, X.: MTD-Net: learning to detect deepfakes images by multi-scale texture difference. IEEE Trans. Inf. Forensics Secur. **16**, 4234–4245 (2021)

Mapping the Intersection of Deep Fake Technology and Consumer Decision-Making: A Comprehensive Bibliometric Analysis

Meenakshi Devi and Veer P. Gangwar

Abstract The purpose of this study is to investigate how deepfake technology and consumer decision-making interact, this study does a thorough bibliometric investigation. Understanding how deepfakes affect customer perceptions, trust, and behavior has become critical for businesses and policymakers as deepfake technology makes its presence felt in the digital world. The relevant 156 publications were extracted from the Scopus database using the keywords "deep fakes" OR "consumer psychology" OR "consumer decision-making" between 2000 and 2024. The analysis was conducted using Vos Viewer and Biblioshiny through R Studio. This study finds important trends, significant authors, and research gaps in the topic by methodically reviewing the current literature. The results highlight how deepfake technology might influence consumer decisions and raise substantial moral questions about truthfulness and fraudulent activity. In summary, the present study offers a substantial perspective on the effects of deepfakes on customer trust and engagement, underscoring the necessity for more investigation in this rapidly developing field.

Keywords Deep fakes · Online shopping · Artificial intelligence · Consumer decision-making · Digital technology · Bibliometric analysis

1 Introduction

In recent years, the practice of shopping online has become increasingly popular due to the widespread use of smartphones and the accessibility of affordable internet connectivity. In addition to benefiting online businesses, this tendency has made

M. Devi (✉) · V. P. Gangwar
Lovely Professional University, Phagwara, Punjab, India
e-mail: meenakshirajput454@gmail.com

V. P. Gangwar
e-mail: gangwarveerdr@gmail.com

S. Taneja et al. (eds.), *Navigating the Deepfake Conundrum: A Manager's Roadmap*, Information Systems Engineering and Management 44,
https://doi.org/10.1007/978-3-031-90742-5_11

the online retail sector more competitive [2]. Businesses spend a lot of money on marketing campaigns to get people to check out their official sites [9]. Companies produce not just goods that fit for taste and visual beauty, hedonic enjoyment, compassion, and several other preferences but are also better and more convenient to utilize and consume [42]. The process by which customers decide which products to buy has been studied by marketers. Purchase-related necessities initiate a series of behaviours that provide value. In the process of trying to meet actual needs, consumers engage in value-seeking behaviours and these requirements drive a consumer's behavior, leading them to seek out appealing means of satisfying their requirements. A value is ultimately produced by the entire procedure, which involves some psychological activities, such as evaluating, experiencing, and behaving [16]. Customers who make purchases online tend to act in relational ways to increase choice-making effectiveness, while minimizing information processing, increasing mental coherence when making selections [15], and lower perceived risk related to forthcoming decisions [19, 20]. Deep fakes are often used on social networking sites and other digital platforms, largely due to how simple and quick they are to make. Developments in artificial intelligence (AI) have made it possible to create believable false audio and video recordings. Deepfakes are made-up pictures, movies, or sounds in which a person's face or voice is substituted [26]. One of the current developments that is causing problems in the media sector to surface is Deepfakes Technologies. These artificial intelligence-generated "exaggerated information" are distributed as real ones [10].

Deep Fakes are producing increasingly appealing fabricated and false representations of reality with higher fidelity and credibility due to various factors such as improved image sensor quality and decision-making, increased data accessibility, and the explosive growth of AI technologies, especially machine learning (ML) and deep neural networks (DNNs) [13]. An average consumer is finding it harder and harder to discern between very realistic counterfeit stuff and the actual one [8]. The use of the internet has surged dramatically [15], with widespread adoption of internet access and a sharp increase in online sales. Researching products and making purchases online has gained immense popularity. Online retailers offer an extensive range of goods, convenient payment and delivery options, and the potential for significant time and cost savings. Online shopping mirrors traditional retail experiences, involving purchases made through a website [17, 37]. If utilized ethically deep fake technology is also beneficial for marketers. Deepfakes seem to be ubiquitous on many online communication platforms such as social media [4], since this process is so easily achieved. The creation of Deepfakes technology creates the opportunity for smart advertising. Integrated consumer tracking solutions are employed by advertising to create commercials and adverts with strong links to the tastes of the common masses or the few. They also can generate, or simply deliver, customized movies for each step of a customer journey. In addition, today, deepfakes are employed to personalize marketing communications and to anthropomorphize material by automatically creating human attributes—faces, voices, and others. Deepfakes should therefore be able to add to the authenticity and effectiveness of the advertising messages [10]. Robust and authentic advertising material is produced via deepfakes and generative adversarial networks, which make use of strong AI creative capabilities. The advertising business will face both possibilities and challenges from altered advertisements when these new tools are embraced more broadly. A variety of professionals including brand executives, advertisement organizers, analysts, creative staff, producers, models, distributors, and regulators will need to consider the possible effects of deceptive marketing on their particular fields [5, 6]. The focus on advertisements is moving toward "synthetic advertising" from content produced and modified only with analog and digital means [24]. Even though synthetic advertising is a more sophisticated kind of altered advertising, it still fits within the larger context. Advertisements created or altered by artificially and automatically producing and altering data are referred to as synthetic advertising. Artificial intelligence (AI) techniques, such as deepfakes and Generative Adversarial Networks (GANs), are typically used in this automated generation of material that presents a highly appealing, although fabricated, version of reality. The now more widely used technique, deepfakes, uses artificial intelligence (AI) to replace a source's characteristics (such as the face, voice, skin tone, and gender) with the traits of an intended audience [5, 6].

In the realm of deep fakes, they might have a hard time knowing whether the person in the video is real or fake, or realizing that experience should be something they're asking. Deepfakes open existing marketing opportunities to deeper engagement across cultural boundaries, improved messaging personalization and effectiveness, and even opportunities for customer-led co-creation, just as (new) hope that deep fakes can be used for the benefit of both customers and business [49]. Artificial Intelligence (AI) brings to bear cutting-edge technology including chatbots, machine learning (Ml), and natural language processing (NLP) to automate and personalize client interactions by producing faster and more effective services. It's a process where you examine the consumer data to get ideas about what they like, what they do, or what they bought. This data helps customize interactions, suggestions, and promotions further and gives clients a more personalized, compelling client journey. Client inquiry and support are made available through continuous accessibility using an AI-driven chatbot or virtual assistant. This practice guarantees the consumer that he will be given support anytime, increasing consumer satisfaction and confidence [34, 41]. As the present paper also demonstrated, the existing body of research suggests that many prior empirical and theoretical works offer insights into deep fake technology meanings and its impact on the consumer decision-making process [29, 38]. However, there is still the need for further systematic examination of this emerging topic to unveil the range of themes and trends evidenced in the current literature on deepfakes. To address this literature gap, in our study, we identified (a) main themes by bibliometric analysis and (b) analysis of the main arguments constituting each theme including authors, journals, keywords, and representing clusters. Considering these gaps, the paper aims to answer the following questions:

1. In what ways does deep fake technology influence consumer decision-making?
2. What are the contexts of the themes identified?
3. What avenues for future deep fake technology and consumer decision-making research are possible?

In this sense, the study aims to actualize the following objectives and can help marketers provide insights into the trajectory of the field's developments, possible obstacles, and implications while outlining future research goals to help explain how and to what extent deep fakes influence consumers' decision-making journey.

2 Literature Review

The Internet has become the information source to go to when customers are making purchasing decisions [39], one of the pivotal steps involves seeking consumer information [18]. To enhance both online and physical sales, businesses must develop a thorough understanding of how customers conduct online information searches [47]. This process typically encompasses several phases, beginning with problem identification, followed by information search, alternative evaluation, and ultimately, the formulation of the purchasing decision [12]. To our understanding, despite the growing adoption of AI by marketing managers and consumers, [27] there has not been a comprehensive study that thoroughly explores the body of knowledge generated on AI within the interconnected realms of psychology, consumer studies, and marketing [46]. Today's modern world faces a threat from photoshopped and edited photographs. One may artificially produce any kind of audio or video clip. It has become more difficult to differentiate manipulated images and videos from their original forms because of artificial intelligence (AI), particularly machine learning (ML) techniques [32]. Conversational software agents, often known as chatbots, are also becoming increasingly popular as a replacement for human chat service providers in modern times. These agents, which frequently use artificial intelligence (AI) methods, are designed expressly to communicate with individuals in natural language. Employing chatbots has shown to be beneficial for businesses, especially when it is associated with a positive consumer experience. The ability of a chatbot to address client difficulties and the relevance of its replies are the key factors determining its efficacy in customer care. Positive outcomes, such as customer happiness, a higher chance of chatbot usage in the future, product purchases, and product suggestions, are usually the result of these characteristics [41].

The AI sector is poised for continued growth and success in the upcoming years. Artificial intelligence (AI) is an incredibly advanced tool with wide-ranging applications across various industries. AI also extends to synthetic media, encompassing autonomously generated material blurring the line between authenticity and falsity [50]. Not all AI programs have malicious intent. Customer-facing AI, such as chatbots and virtual assistants, is already used. However, the reception of deep fake technology, which can convincingly mimic "real" video using AI, remains uncertain [49]. The term "deepfakes" denotes a form of digitally manipulated content characterized by the creation of highly realistic fake material through the manipulation of images and/or videos using artificial intelligence (AI) algorithms [23]. Deepfakes involve portraying individuals who were not originally present in the source video or image, often deceiving viewers by depicting them engaging in actions or expressing sentiments that they have not done or conveyed. This technology can be employed in advertisements to falsely portray individuals endorsing products they may have never encountered, leading consumers to mistakenly believe in their genuine use or recommendation [22].

3 Deep Fake Technology and Consumers

Online commerce has experienced significant growth over the past few decades and has established itself as an ongoing trend for a variety of goods [40]. The purchasing behavior of consumers is multifaceted, impacted by several variables, and involves multiple phases [14]. Numerous academics made an effort to create models that capture the complexity of buyer decision-making while describing the process of making decisions [43, 52]. This process was generally shown in these models as a multistage process with five stages. These steps involve identifying the issue, looking for information before making a purchase, analyzing your options, making a buy, and evaluating the post-purchase experiences [1]. In the last decades, experts studying marketing have started to look at these problems from the perspective of the customer [25]. Although emotional expressions play a crucial role in human communication and significantly impact the way people perceive misinformation [45] so, it is crucial to understand the potential of deep fakes to accurately portray customers' emotional expressions [3]. The relationship between emotion and cognition is recognized from a psychological standpoint [21]. There are many possible uses for this technology, including media manipulation, evidence fabrication, and personal sabotage. The emergence of synthetic technology is concerning as it is simple to employ and incredibly lifelike, making it hard to identify [30].

The widespread use of deepfake media to disseminate misleading information, sway public sentiment, and hurt people has made the detection of such materials a hot study topic. Artificial intelligence has brought in a big new technological advancement in the form of deepfakes: Hyper-realistic synthetic media content created by an algorithm [35]. The vast reach and speed of online platforms are fueling a Trojan horse of sorts that can deceive millions and cause wide-magnitude marketplace fraud. Deepfakes are fake videos, fake images, and fake sound clips of people saying or doing things they never have—in reality. It's a deliberate attempt to bring people in, but that's a threat to consumers and businesses. Nevertheless, commercialization of this technology is also lucrative, given both increased consumer activity in virtual environments and the current value of developing such immersive experiences. This implies that firms will put more time and effort into exploring the advantages of deep fakes [33].

4 Challenges of Deep Fakes

Deepfakes pose a danger to our political structure, culture, and economy since they (1) make it harder to distinguish between true and false news (2) erode public confidence in government information and consumer purchase intention [36] and (3) generate concerns about cyber security for individuals and institutions. Deepfakes will likely cause a significant problem for the broadcasting sector in terms of trust among customers. Because deepfakes are more difficult to identify and consumers are more likely to think they are real, they present a bigger threat than "traditional" fake news. Obtaining access to witness-shot video footage can give a news media outlet a competitive edge, but fake material carries a higher risk of fraud [48]. Furthermore, deepfakes can deceive customers and raise market volatility, which might lead to their psychological discomfort and mistrust of companies. This might weaken customers' intentions to make purchases [31]. Deepfakes may contribute to customer mistrust of firms and psychological pain by raising ambiguity in markets and deceiving consumers. When consumers undergo this pain, they feel nervous, unsatisfied, and negatively stimulated by advertisements. It is made up of unpleasant or unsettling elements like anger, fear, or perplexity. Customers have a bad opinion of advertising on many online and physical channels when they are irritated [8].

5 Benefits of Deepfake Technology

In addition to these uses, deepfake technology can also be used positively in many other industries including movies, educational media and digital communications, games and entertainment, social media and healthcare, material science, and many business fields, such as fashion and e-commerce. There is a business interest in applying deepfake technology into the shoulders of the brand, as it would radically change e-commerce and advertising. For instance, brands may take a supermodel and have them wear fashion clothes on a wide array of models that are, well, not a supermodel, with different skin tones, heights, weights, and even body types. Additionally, deep fakes let users go very personal. The technology also creates targeted fashion ads based on time, weather, and viewer, turning consumers themselves into models via content that previews how an outfit would look on them without owning it. Suppose the technology could turn people into digital avatars of themselves and send them along to shop online. In that case, that sounds like an obvious use case, moving between e-stores without ever having to take clothes off the hanger or put on a change of clothes. It's also about being able to try on a bridal gown or suit in digital form and then virtually checking out a wedding venue. Also, AI can offer unique artificial voices that can make it easier to brand companies and products [48].

Figure 1 illustrates the trends in topics covered in previous studies on deepfake technology and its impact on consumer decision-making. The figure shows that previous studies have primarily focused on a few specific issues related to the effects

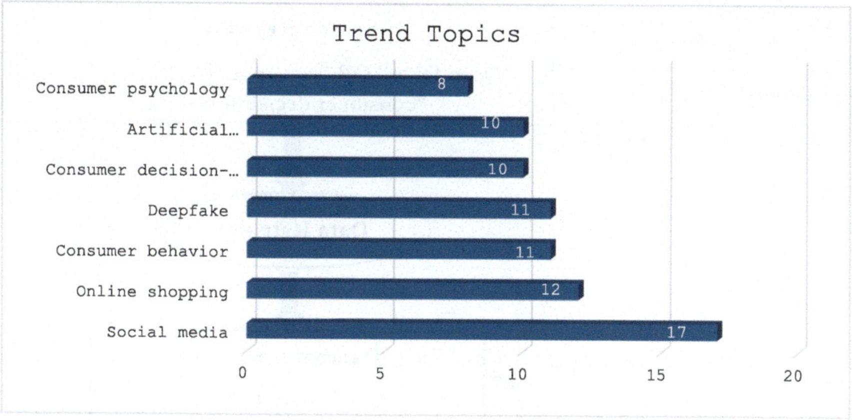

Fig. 1 Trend topics (*Source* Author's compilation)

of deepfakes on consumer decisions from the years 2000 to 2024. The extensive literature review helped create Fig. 1, which showcases the main topics of interest in recent years. "Social media" emerges as the most prominent subject, with a frequency of 17, highlighting its crucial role in deepfake technology and its influence on consumer behavior through digital platforms. Topics such as "Online shopping," "Deepfake," and "Consumer behavior," each with frequencies of 12 and 11, have been the focus of both past and present studies, emphasizing the importance for marketers to understand consumer behavior in a rapidly evolving technological landscape. On the other hand, "Artificial intelligence," "consumer decision-making," and "consumer psychology," each with frequencies of 10 and 8, are also significant but hold a secondary position in this discussion. Collectively, these topics provide insights into how consumers interact in online environments, make purchasing decisions, and navigate the era of deepfakes. In conclusion, the research emphasizes the critical importance of understanding customer behavior, decision-making processes, and marketing strategies in the context of consumer purchasing decisions and deepfake technology.

6 Methodology

In this research (Fig. 2), we utilized bibliometric analysis to comprehensively evaluate academic literature. This method is widely employed to track conceptual development and identify key elements such as countries, authors, and journals across various fields of study. It involves a systematic review of previously published studies to identify patterns, highlight statistical findings, make comparisons, identify knowledge gaps, and propose future research directions [44]. We specifically filtered English language publications from the Scopus database using the search terms "deep fakes,"

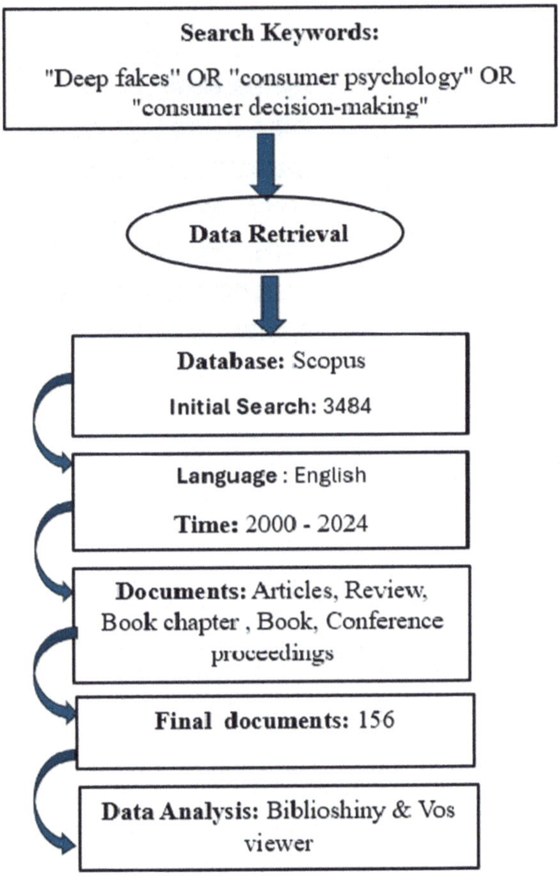

Fig. 2 Procedure for selecting relevant articles (*Source* Authors compilation)

"consumer psychology," and "consumer decision-making," and included research articles, book chapters, books, conference papers, and journal reviews published between 2000 and 2024. Although both Web of Science and Scopus are widely used bibliometric databases, we chose the Scopus database due to its broader coverage [28]. For our bibliometric analysis, we employed the R-based bibliometrix program (biblioshiny) and VOS viewer.

7 Results

The current analysis bases its findings on 156 articles that were published in different scholarly publications between 2000 and 2024 discussing consumer decision-making and deepfake technology. The data presented in Fig. 3 shows the annual article production from 2015 to 2024. In 2015 and 2016, only 2 articles were published each

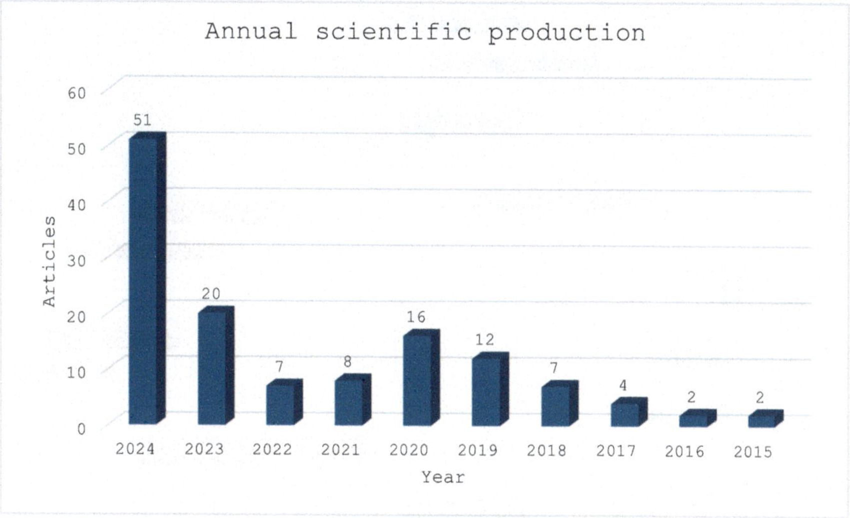

Fig. 3 Annual scientific production (*Source* Author's compilation)

year, indicating minimal research in the area of deepfakes during that time. However, starting from 2017, there was a noticeable increase in the number of publications. The following years displayed fluctuations in the number of articles, with 2021 and 2022 showing a significant decrease to 8 and 7 articles, respectively. There was a consistent increase in articles in 2023 (20) and 2024 (51). Overall, the data highlights the variations in the number of articles and a clear shift towards more contemporary and impactful research.

The following relationships can be seen between the various terms that other researchers have investigated when it comes to deep fakes and consumers' choices, as depicted in Fig. 4. The figure shows the five clusters represented in (C1) red, (C2) blue, (C3) dark green, (C4) yellow, and (C5) light green. The link words in C1 and C2 are consumer behavior and e-commerce, related to other terms of other clusters such as social media and deep fake social -media and deep fake are also central terms in C3 connected with machine learning, online shopping, e-commerce, consumer psychology and marketing. This means that deep fake is a very important factor that helps to determine the behaviour of consumers on the internet. C4 and C5 represent the smallest cluster which indicates that the majority of the authors used the terms; "consumer", "marketing", "machine learning" and "online reviews" to analyse the psychological aspect of the consumer's online reviews" [40] is a mostly used keyword by the authors to understand the psyche of the consumers.

The connections between different keywords concerning deepfakes and consumer behaviour are depicted in the form of a diagram.

This section compared a list of ten documents and their citation indexes at a world level. Using Biblioshiny software, we found the most cited documents in the different fields (Table 1). The analysis showed that the work by Häubl G. in

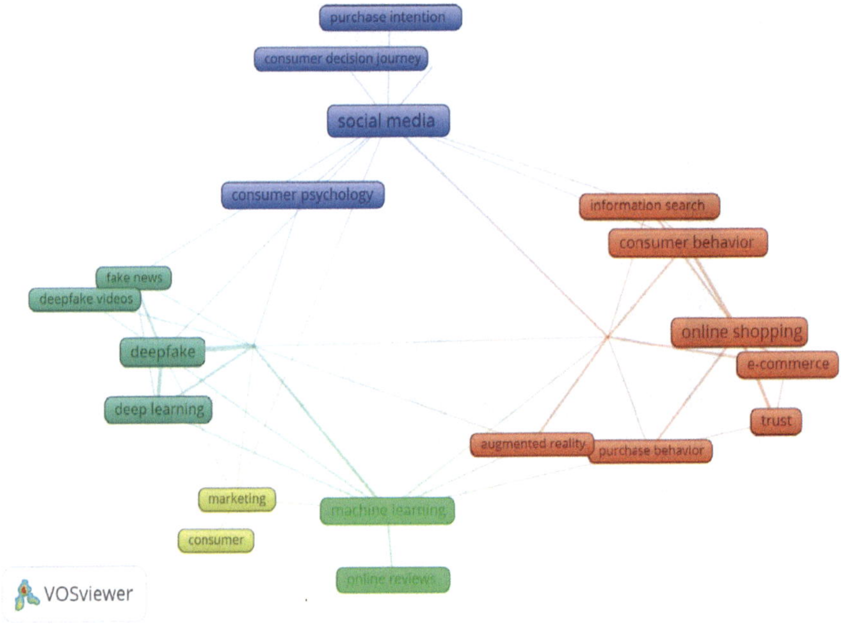

Fig. 4 Co-occurrence of author keywords (*Source* Vos viewer)

"Marketing Science" from 2000 has the highest overall citation index of 1130, which also speaks about the research and practical activity of the marketer in the academic community. Next is Westerlund M's paper published in Technovation in 2019 with 440 citations while the next work is Shaw D published in the European Journal of Marketing in 2003 with 227 citations. That is why these high citation counts demonstrate that they are the most productive authors who have contributed to the area that has to do with deep fake technology, consumer buying decisions, artificial intelligence, and consumer behaviour. In Table 1, we also have papers by Mishra R. in International Journal of Consumer Studies (2021), Goodrich K. in Journal of Market Communication, and Utz S. in Electronic Commerce Research and Applications in 2014 and 2012 respectively which have been cited 206 times. Some other authors who also contribute are Hsu M.-H. (2014), Nambisan P. (2011), Maity M. (2014), and Umit Kucuk S. (2007) with a citation of 187, 179, 144, and 143, respectively.

Such information shows the role of the documents within their academic subject or domain while embodying the significance and impact of those documents throughout the world.

Table 2 reveals how the various academic journals have made an impact in their respective field. Thus, concerns like "Journal of Business Research" and "International Journal of Retail and Distribution Management" can be considered to make a significant impact with the h-index of 4 and 3 respectively, this implies that a minimum of 3 articles published by such concerns have been cited at least three times proving sturdy and important influence. An m-index of 0.2 and 0.273 supports

Table 1 Most global cited documents

Documents	DOI	Total citations	TC per year
Häubl G., 2000, Marketing Science	https://doi.org/10.1287/mksc.19.1.4.15178	1130	45.2
Westerlund M., 2019, Technology Innovation Management Review	https://doi.org/10.22215/TIMREVIEW/1282	440	73.33
Shaw D., 2003, European Journal of Marketing	https://doi.org/10.1108/03090560310487202	227	10.32
Mishra R., 2021, International Journal of Consumer Studies	https://doi.org/10.1111/ijcs.12617	224	56
Goodrich K., 2014, Journal of Market Communication	https://doi.org/10.1080/13527266.2013.797773	206	18.73
Utz S., 2012, Electronic Commerce Research and Applications	https://doi.org/10.1016/j.elerap.2011.07.010	206	15.85
Hsu M.-H., 2014, Internet Research	https://doi.org/10.1108/IntR-01-2013-0007	187	17
Nambisan P., 2011, Journal of Business Research	https://doi.org/10.1016/j.jbusres.2010.09.006	179	12.79
Maity M., 2014, Decision Support System	https://doi.org/10.1016/j.dss.2014.01.008	144	13.09
Umit Kucuk S., 2007, Technovation	https://doi.org/10.1016/j.technovation.2006.05.002	143	7.94

Source Biblioshiny

further that the researcher has been establishing a positive and continuous contribution in the field. These journals have a big total citation which is 472 for the first one and 154 for the second one they are in a crucial place. It can also be seen that only two of them had a remarkable impact with having h-indices 2, these are the "Computers in Human Behavior" and the "Decision Support Systems". The above journals in particular have total citations of 149 and 146 respectively proving their impact on consumer studies. The Scopus Indexed was available in the "Electronic Commerce Research and Applications" and "International Journal of Consumer Studies". The total citations are 253 and 238 respectively which confirms the potential to have a higher impact in the future.

In sum, the data gives a clear understanding of the effects produced by these journals. This includes citation records, publication date, and their production within various fields of specialty.

After analyzing 156 cited papers using Vos Viewer software, we can conclude that the lines connecting the circles in the diagram represent the associations between countries. The thickness of the lines indicates the strength of the association. The diagram (Fig. 5) helped us identify essential countries and their relationships. For example, the United Kingdom is associated with most countries, such as France, the

Table 2 Sources local impact

Element	H_Index	G_Index	M_Index	TC
Journal of Business Research	4	4	0.2	472
International Journal of Retail and Distribution Management	3	3	0.273	154
Computers in Human Behavior	2	2	0.222	149
Decision Support Systems	2	2	0.182	146
Electronic Commerce Research and Applications	2	5	0.154	253
International Journal of Consumer Studies	2	2	0.5	238
Internet Research	2	2	0.182	207
Journal of Consumer Psychology	2	3	0.095	103
Journal of Marketing Communications	2	2	0.182	238
Qualitative Market Research	2	3	0.133	43

Source Biblioshiny

United States, India, Australia, and Canada, indicating its significant role in international scientific collaboration. Similarly, the United States is connected to many other countries, including China, India, the United Kingdom, and Canada, showing its importance in collaboration for publication. India is also a rapidly growing player and is associated with many countries, including the United Kingdom, France, the United States, and Pakistan.

Overall, the diagram provides valuable insights into how nations cooperate on research, illustrating the trend of worldwide scientific collaboration.

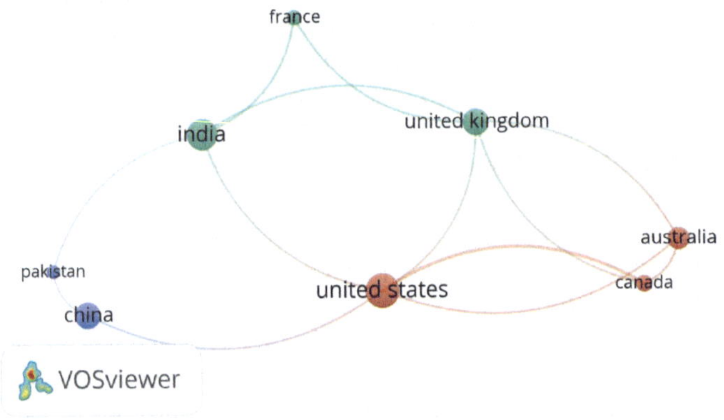

Fig. 5 Countries collaboration (*Source* Vos Viewer)

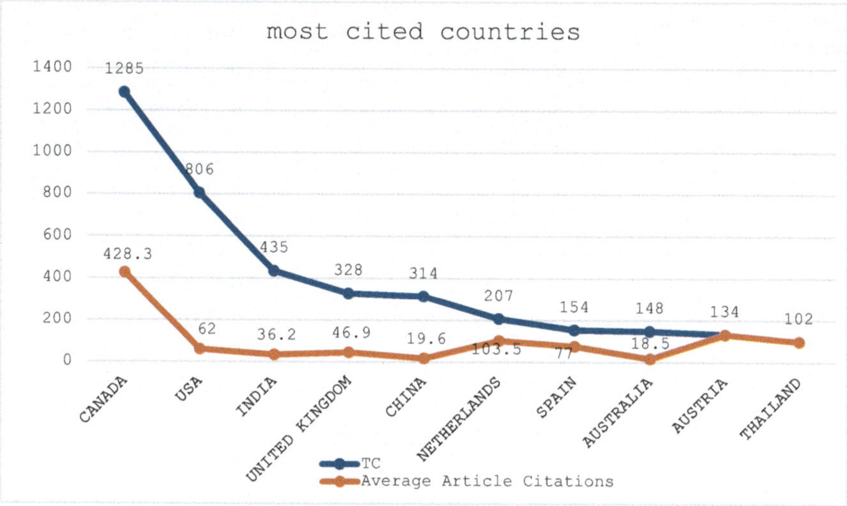

Fig. 6 Most cited countries (Author's compilation)

This section compared a list of the top 10 counties and their global citation counts. Using Biblioshiny software, the top cited countries from 2000 to 2024 by field were determined (Fig. 6). It was also seen that Canada has the highest total citation of 1285 and an average article citation of 428.3 having a strong impact on each paper produced about the deep fake technology and consumer behaviour. Second on is the USA which also reflects a big hit with a total of 806 and an average article citation of 62. The third cited country is India with an average citation of 36.2 and a total citation of 435; the results reveal a good potential in the fields of deep fakes, consumer decision-making buying process, and other domains. The other countries that also extend great contributions and which are often most cited include the United Kingdom, China, Netherlands, Spain, Australia, Austria, and Thailand.

This is evident when analyzing the data showing that other countries also contribute and all these countries across the fields have made remarkable and diverse contributions to scholarly production.

The data in Table 3 includes the academic articles written by these authors and focuses on the number of articles published and fractional citation credit taken because of co-authorship. Kietzmann J. has been computed to have fractionalized contribution for the following 3 articles with the score 0.7333. Ahuja V. and Jain N. co-authored two contributing articles with fractionalized scores of 0.6667. As to authorship, Bačík R., Gburová J., and Gonçalves H.M. contributed to two articles each and received full fractionalized credit of 1, which indicates sole or primary authorship. In the same way, Kumar D. was involved in 2 articles, concretely, he contributed 0.5 which shows that he is a co-author of the articles. Letheren K. was a slightly shared author in 2 articles with a score of 0.5333. Last, Li X. wrote 2 articles and Li Y. 3 articles; both have obviously contributed to co-authored articles since the

Table 3 Most relevant authors

Authors	Articles	Articles fractionalized
Kietzmann J.	3	0.73333333
Ahuja V.	2	0.66666667
Bačík R.	2	1
Gburová J.	2	1
Gonçalves H.M.	2	1
Jain N.	2	0.66666667
Kumar D.	2	0.5
Letheren K.	2	0.53333333
Li X.	2	0.7
Li Y.	2	0.83333333

Source Author's compilation

scores are fractionalized—0.7 and 0.8333, respectively. This fractionalized metric gives the opposite and transparent impression of the researcher's productivity in collaborative work. Table 3 provides an overview of the number of articles authored by various authors across different institutions and countries.

Figure 7 displays the results of the top 10 most relevant sources that have published research articles in the areas of "deep fakes and consumer decision-making." The outcome is based on Scopus information obtained from 2000 and beyond. It is mentioned that "Electronic Commerce Research and Applications" is the most relevant source, with 5 articles, which shows that this journal is active in publishing articles related to the areas of consumer behaviour and deepfake technology. Other relevant sources include "Heliyon" and "Journal of Business Research," with four articles each. The Journals with 3 articles each are "Cogent Business and Management", "International Journal of Retail and Distribution Management", "Journal of Consumer Psychology", and "Qualitative Market Research". "Artificial Intelligence Review", "Computers in Human Behavior", and "Behavioral Sciences" show that these journals have fewer publications in the areas of deep fake technology and consumer decision-making during online purchases i.e. 2 articles each. From the given data it can be concluded that "Electronic Commerce Research and Applications" falls topmost among the top 10 journals.

The thematic map shown in Fig. 8 offers the findings from the systematic literature review that was conducted here solely centering "deepfake technology" and "consumer decision-making" studies. This map is divided into four distinct sections: These include niche themes, declining themes, motor themes, and finally the basic themes. Each of these sections is divided into sub-sections based on some common themes. Hence, the "niche themes" segment contains some of the most important themes, for instance, consumer psychology and consumer decision-making. On the other hand, the 'declining/emerging themes' look into areas of concern such as social networks and consumer decision making which earlier posed so much concern but has or is fading out. The section called "motor themes" includes important topics

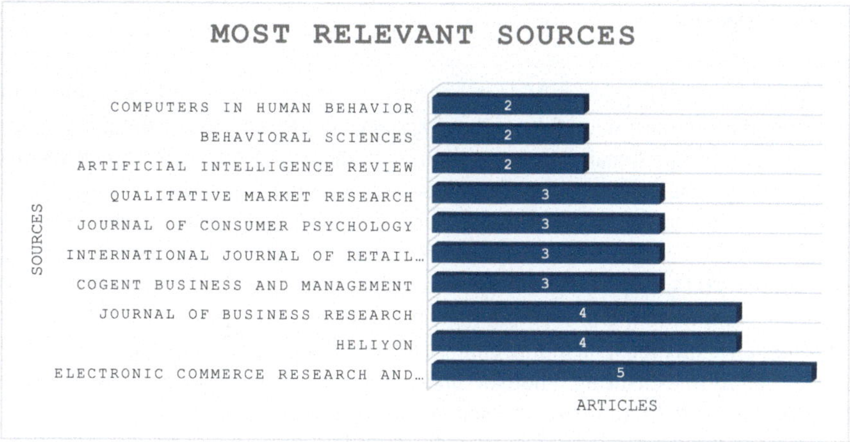

Fig. 7 Most relevant sources (*Source* Author's compilation)

such as deepfake and AI. The broader "basic themes" include buying and consumption over the internet and consumer behaviour which are core aspects of consumer research. This thematic map also presents the current major lines of interest and research studies related to consumers' behavior in the context of a deepfake product or advertising, thus illustrating the fluidity of the field.

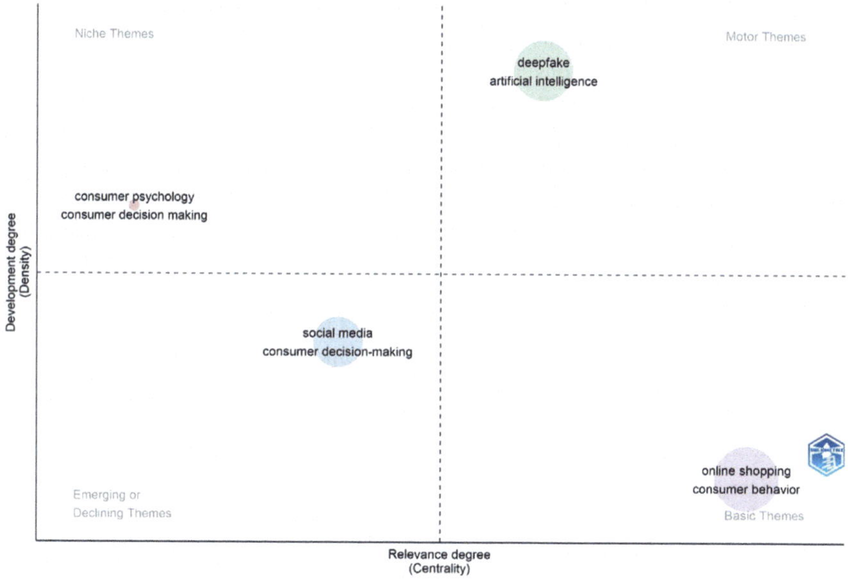

Fig. 8 Thematic map (*Source* Biblioshiny)

8 Discussion

Our recent study unequivocally demonstrates the effectiveness of bibliometrics in exploring the realm of deep fake technology and its impact on consumer decision-making. Our comprehensive findings illuminate the interdisciplinary nature of this field of study. The primary aim of our research is to meticulously track the evolution of deep fake technology and consumer decision-making over the past 25 years. To achieve this, we rigorously analyzed 156 publications from 122 sources spanning from 2000 to 2024, employing cutting-edge Biblioshiny and Vos Viewer software for thorough examination. Our methodological approach encompassed a diverse range of bibliometric methods, including citation analysis, co-occurrence of keywords, and thematic analyses, ensuring a holistic understanding of the subject matter. Furthermore, our in-depth literature review shed light on emerging e-commerce trends [14], emphasizing the integration of social media, mobile commerce, and video marketing to enhance customer retention and drive sales. The proliferation of smartphones has significantly propelled the growth of mobile commerce, despite challenges posed by the unethical use of deep fakes by advertising companies, leading to feelings of deception among consumers [13]. Additionally, the study includes a detailed analysis that offers valuable insights into the interconnectedness of ideas and concepts within the research domain. This systematic study has significantly contributed to our understanding of the concept of deep fakes and consumer psychology. Considering that consumers hold significant power in the market, marketers must understand their thought processes, proactively address any issues they encounter while shopping online, and ultimately enhance customer satisfaction. To begin, we have thoroughly outlined the field using various robust tools, including thematic maps, influential authors, institutional affiliations, and relevant productive journals.

Consumer decisions are at the vane of the development of deepfake technology. Artificial intelligence-produced (or 'deliberately spoofed') media (deepfakes) are media that combine real and fake materials, perhaps fooling consumers and creating doubt about the sources of information on which they base their confidence. This technology may have a great impact on consumer behaviour by producing incredibly commercial ads, phony celebrity endorsements, or even custom-fit messaging [11]. It can also make customers think they are making decisions based on lies. As deepfake material becomes more popular, traditional indicia like visual authenticity, testimonials and even influencer endorsements that customers rely on to make smart decisions are becoming less reliable. They are, however, equally well suited to brands and marketers looking to deploy deepfake technology in constructive, rather than destructive, ways—for instance, to deliver personal and immersive consumer experiences that drive engagement. The main challenge is to make this technology transparent and ethical use [52]. Authenticity is a priority for brands, so they need to have mechanisms that help verify the legitimacy of digital content for users to believe what they are reading. Deepfake research needs to center on how deepfake exposure influences consumer trust in the future, and what regulatory frameworks and technological solutions (such as blockchain verification and AI detection tools) could be developed to curb such harms. Increasingly, the balance of newness and consumer protection will determine the part this technology will play in marketing and business decision making as deepfakes evolve.

9 Managerial Implications

The bibliometric analysis highlights that our research provides a comprehensive understanding of deep fakes from the consumer perspective. This insight will enable businesses to identify valuable sources for real-world insights into this concept. For product-related media, managers need to implement strict verification procedures and transparency policies to preserve customer confidence and protect the value of the brand. A brand's reputation might be harmed by the misuse of deepfake technology, thus it's critical to invest in deepfake detection tools and encourage openness. To improve customer connections, businesses should inform customers about the moral applications of deepfake technology. For e-commerce companies, the growing emphasis on consumer value has significant management implications. This involves ensuring that the entire shopping experience, including customer service and website usability, consistently surpasses the value expectations of the target audience. Prioritizing value allows companies to enhance customer satisfaction, foster loyalty, and secure a unique position in a competitive online market.

10 Future Lines of Research

Further investigation into the relationship between fake ads and customer behavior, as well as how it affects brand experiences and satisfaction, may be required in future studies. Scholars might also look at the mediated effects of utilitarian and hedonistic values. Furthermore, it's unclear how consumers will react to hyper-personalized advertisements just yet. How will customers balance the ease and usefulness of customization against the perceived invasion of their privacy when advertisers produce ever more individualized ads for every individual? The alteration of advertisements is getting more complex, and this gives scholars and practitioners a better understanding of the possibilities and possible problems it presents. The concentration of future research on deepfake technology and consumer decision-making should be how to unveil the psychological and behavioral consequences of deepfake exposure, and how to mitigate or detect such exposure. Additionally, research should be directed towards how technology solutions are helping consumers trust and managers do their best in fighting the misinformation. However, as deepfake technology continues to expand, it's going to become even more important to leverage interdisciplinarity in psychology, marketing, and computer science to develop sturdy frameworks for the ethical and productive use of synthetic media in consumer markets.

11 Conclusion and Limitations of the Study

The present study identifies distinct periods within this domain, with the years 2019 to 2020 exhibiting relatively stagnant trends, while the subsequent period from 2023 onwards witnessed a consistent annual increase in research output. Remarkably, the journal "Journal of Business Research" emerged as the foremost source of papers on this subject. Through meticulous bibliometric analysis, we have also identified critical gaps in the existing body of research. Finally, this study uncovers the profound ramifications of the deepfake technology on consumers' purchase contemplations and the branding procedure, reviewing both the positive and negative parts of synthetic media in the promotion. Deepfakes are advancing at an increasing pace and increasing availability, and as such marketers are opening up new ways that can be leveraged to create more personalized, interactive content to boost consumer engagement, as well as streamline customer service and add new levels of immersion to brand storytelling. However, as with most technologies, whether this raises the issue of misinformation, trust, or brand credibility, it also has ethical concerns associated with it. Deepfakes have the potential to obscure the lines that consumers rely heavily on when making purchasing decisions—perceived authenticity—and when used incorrectly, these misused deepfakes inspire distrust and suspicion. As this is now well into the future, brands must employ a strong ethical foundation around deepfake technology and engage with practices that are transparent and explicitly tell consumers that synthetic media is being used and are actively developing methods like blockchain or AI-based detection technologies that will keep content authentic. At the same time, brands can protect their image from potential harm and cultivate loyalty in conditions where trust is becoming the key factor and consumers demand more and more information about the company and product provenance [7]. Moreover, creating awareness about exhibiting deepfake in general and ethical uses thereof, so that consumers can make sound decisions, strengthens the brand's side and safeguards consumers. Therefore, this study calls for caution in adopting deepfake technology by bridging the gap whereby consumers embrace the technology responsibly to build good relationships with them in the dynamically changing world.

Despite the insights generated by this study as to the managerial effects of deepfake technology, this research has its limitations. As for the study is concerned the data is only taken from the Scopus database. To get more effective results data should also be extracted from other databases such as Web of Science, Google Scholar, etc. The relationship between fake ads and customer behavior, as well as its impact on brand experiences and satisfaction, may need further investigation. Scholars could also explore the effects of utilitarian and hedonistic values. Additionally, the reaction of consumers to hyper-personalized advertisements and the balance between customization and privacy invasion is still unclear. The increasing complexity of ad alterations provides scholars and practitioners with a better understanding of the possibilities and potential issues it brings. The other limitation is as the deepfake technique emerged quite recently, the consumers' recognition and understanding of deepfakes can be low and differ across segments thus potentially affecting the study. Furthermore, this research mainly involves imagining situations, and the results might not reflect the variable responses of consumers to real deepfake content. Future studies should consider these limitations in their research attempts.

References

1. Akalamkam, K., Mitra, J.K.: Consumer pre-purchase search in online shopping: role of offline and online information sources. Bus. Perspect. Res. **6**(1), 42–60 (2018). https://doi.org/10.1177/2278533717730448
2. Ali, D.M., Amir, D.H.: Understanding consumer herding behavior in online purchases and its implications for online retailers and marketers. Electron. Commer. Res. Appl. **64**, 1–39 (2024). https://doi.org/10.1016/j.elerap.2024.101356
3. Becker, C., Conduit, R., Chouinard, P.A., Laycock, R.: Can deepfakes be used to study emotion perception? A comparison of dynamic face stimuli. Behav. Res. Methods **56**(7), 7674–7690 (2024). https://doi.org/10.3758/s13428-024-02443-y
4. Bhardwaj, S., Kumar, N., Gupta, R., Baber, H., Venkatesh, A.: How social media influencers impact consumer behaviour? Syst. Lit. Rev. Vis. 1–16. https://doi.org/10.1177/09722629241237394
5. Campbell, C., Plangger, K., Sands, S., Kietzmann, J.: Preparing for an era of deepfakes and AI-generated ads: a framework for understanding responses to manipulated advertising. J. Advert. **51**(1), 22–38 (2022). https://doi.org/10.1080/00913367.2021.1909515
6. Campbell, C., Plangger, K., Sands, S., Kietzmann, J., Bates, K.: How deepfakes and artificial intelligence could reshape the advertising industry the coming reality of AI fakes and their potential impact on consumer behavior. J. Advert. Res. **62**(3), 241–251 (2022). https://doi.org/10.2501/JAR-2022-017
7. Chang, T.P.V., Rhodes, J., Lok, P.: The mediating effect of brand trust between online customer reviews and willingness to buy. J. Electron. Commer. Organ. **11**(1), 22–42 (2013). https://doi.org/10.4018/jeco.2013010102
8. Chen, Y.: Exploring the Effect of Deepfake-Based Advertising on Consumers' Attitudes and Behaviors (2024)
9. Dheepalakshmi, S., Geetha, R.: Exploring the effect of marketing factors on customers' online purchase intention using the stimulus organism response (SOR) framework. Int. Res. J. Multidiscip. Scope **5**(2), 334–344 (2024). https://doi.org/10.47857/irjms.2024.v05i02.0481
10. Dimuthu Maduranga Arachchi, H.A., Samarasinghe, G.D.: Impact of deepfake advertising attributes on consumers' hedonic & utilitarian values and brand credibility. In: Proceedings—International Research Conference on Smart Computing and Systems Engineering, SCSE 2024, June (2024). https://doi.org/10.1109/SCSE61872.2024.10550666
11. Durmishi, A., Durmishi, L.: The impact of influencer marketing on the consumer decision-making process. Multidiscip. Sci. J. **6**(11), 2024227 (2024). https://doi.org/10.31893/multiscience.2024227
12. Etco, M., Sénécal, S., Léger, P.M., Fredette, M.: The influence of online search behavior on consumers' decision-making heuristics. J. Comput. Inf. Syst. **57**(4), 344–352 (2017). https://doi.org/10.1080/08874417.2016.1232987
13. Gambín, Á.F., Yazidi, A., Vasilakos, A., Haugerud, H., Djenouri, Y.: Deepfakes: current and future trends. Artif. Intell. Rev. (Springer Netherlands) **57**(3) (2024). https://doi.org/10.1007/s10462-023-10679-x
14. Handoyo, S.: Purchasing in the digital age: a meta-analytical perspective on trust, risk, security, and e-WOM in e-commerce. Heliyon **10**(8), e29714 (2024). https://doi.org/10.1016/j.heliyon.2024.e29714
15. Hansen, H., Melbye, E.L.: Negative information, cognitive load, and taste perceptions. J. Food Products Market. **26**(3), 185–196 (2020). https://doi.org/10.1080/10454446.2020.1740129

16. Hazari, S., Bergiel, B.J., Sethna, B.N.: Hedonic and utilitarian use of user-generated content on online shopping websites. J. Mark. Commun. **23**(6), 572–591 (2017). https://doi.org/10.1080/13527266.2016.1143383
17. Helmi, A., Komaladewi, R., Sarasi, V., Yolanda, L.: Characterizing young consumer online shopping style: Indonesian evidence. Sustainability (Switzerland) **15**(5) (2023). https://doi.org/10.3390/su15053988
18. Houdek, P., Koblovský, P., Šťastný, D., Vranka, M.: Consumer decision making in the information age. Society **55**(5), 422–429 (2018). https://doi.org/10.1007/s12115-018-0283-5
19. Hsu, M.H., Chuang, L.W., Hsu, C.S.: Understanding online shopping intention: the roles of four types of trust and their antecedents. Internet Res. **24**(3), 332–352 (2014). https://doi.org/10.1108/IntR-01-2013-0007
20. Imaniya, T., Agus, A.A.: The influence of consumer's decision making process to online purchase behavior analysis. In: Proceedings—2019 2nd International Conference of Computer and Informatics Engineering: Artificial Intelligence Roles in Industrial Revolution 4.0, IC2IE 2019, pp. 116–121 (2019). https://doi.org/10.1109/IC2IE47452.2019.8940898
21. Iordanova, E., Stainton, H.: Cognition, emotion and trust: a comparative analysis of Cambodia's perceived and projected online image. Tour. Stud. **19**(4), 496–519 (2019). https://doi.org/10.1177/1468797619837970
22. Karpinska-Krakowiak, M., Eisend, M.: Realistic portrayals of untrue information: the effects of deepfaked ads and different types of disclosures. J. Advert. 1–11 (2024). https://doi.org/10.1080/00913367.2024.2306415
23. Kaur, A., Noori Hoshyar, A., Saikrishna, V., Firmin, S., Xia, F.: Deepfake video detection: challenges and opportunities. Artif. Intell. Rev. **57**(6). Springer Netherlands (2024). https://doi.org/10.1007/s10462-024-10810-6
24. Kietzmann, J., Mills, A.J., Plangger, K.: Deepfakes: perspectives on the future "reality" of advertising and branding. Int. J. Advert. **40**(3), 473–485 (2021). https://doi.org/10.1080/02650487.2020.1834211
25. Kim, T.W., Usman, U., Garvey, A., Duhachek, A.: Artificial intelligence in marketing and consumer behavior research. Found. Trends in Market. **18**(1) (2023). https://doi.org/10.1561/1700000078
26. Mahmud, B.U., Sharmin, A.: Deep insights of deepfake technology: a review. **5**, 13–23 (2021). http://arxiv.org/abs/2105.00192
27. Mariani, M.M., Perez-Vega, R., Wirtz, J.: AI in marketing, consumer research and psychology: a systematic literature review and research agenda. Psychol. Mark. **39**(4), 755–776 (2022). https://doi.org/10.1002/mar.21619
28. Mishra, H.G., Pandita, S., Bhat, A.A., Mishra, R.K., Sharma, S.: Tourism and carbon emissions: a bibliometric review of the last three decades: 1990–2021. Tour. Rev. **77**(2), 636–658 (2022). https://doi.org/10.1108/TR-07-2021-0310
29. Morales, C., VIzcaino, A.J.: Mercados y Negocios and food security. Scielo **21**, 1–26 (2022). https://www.scielo.org.mx/scielo.php?pid=S2594-01632020000200093&script=sci_arttext
30. Nas, E., de Kleijn, R.: Conspiracy thinking and social media use are associated with ability to detect deepfakes. Telemat. Inform. **87**(November 2023), 102093 (2024). https://doi.org/10.1016/j.tele.2023.102093
31. Nguyen, N.Q., Nguyen, H.L., Trinh, T.G.: The impact of online and offline experiences on the repurchase intention and word of mouth of women's fashion products with the intermediate trust factor. Cogent Bus. Manag. **11**(1) (2024). https://doi.org/10.1080/23311975.2024.2322780
32. Qadir, A., Mahum, R., El-Meligy, M.A., Ragab, A.E., AlSalman, A., Awais, M.: An efficient deepfake video detection using robust deep learning. Heliyon **10**(5), e25757 (2024). https://doi.org/10.1016/j.heliyon.2024.e25757
33. Rahman, A., Mustak, M., Salminen, J., Matti, M.: Deepfakes: deceptions, mitigations, and opportunities. J. Bus. Res. **154**(September 2022) (2023)
34. Ramanathan, U., Williams, N.L., Zhang, M., Sa-nguanjin, P., Garza-reyes, J.A., Borges, L.A.: A new perspective of e-trust in the era of social media: insights from customer satisfaction data. In: IEEE Transactions on Engineering Management, vol. 69, no. 4, pp. 1417–1431 (2022). https://doi.org/10.1109/TEM.2020.2985379

35. Rana, P., Bansal, S.: Exploring deepfake detection: techniques, datasets and challenges. Int. J. Comput. Digit. Syst. **15**(1), 769–781 (2024). https://doi.org/10.12785/ijcds/160156
36. Ruiz-Mafe, C., Chatzipanagiotou, K., Curras-Perez, R.: The role of emotions and conflicting online reviews on consumers' purchase intentions. J. Bus. Res. **89**(February), 336–344 (2018). https://doi.org/10.1016/j.jbusres.2018.01.027
37. Saha, S., Sharma, A.: The impact of consumer decision making styles on the purchase intention of branded apparels online. Int. J. Sci. Technol. Res. **9**(4), 449–453 (2020)
38. Schuster, L., Drennan, J., Lings, I.: Understanding consumers' decisions to adopt technology-enabled transformative services. Serv. Ind. J. **35**(15–16), 846–864 (2015). https://doi.org/10.1080/02642069.2015.1090979
39. Senecal, S., Kalczynski, P.J., Nantel, J.: Consumers' decision-making process and their online shopping behavior: a clickstream analysis. J. Bus. Res. **58**(11 SPEC. ISS.), 1599–1608 (2005). https://doi.org/10.1016/j.jbusres.2004.06.003
40. Sharma, P., Ueno, A., Dennis, C., Turan, C.P.: Emerging digital technologies and consumer decision-making in retail sector: towards an integrative conceptual framework. Comput. Hum. Behav. **148**. https://doi.org/10.1016/j.chb.2023.107913
41. Singh, P., Singh, V.: The power of AI: enhancing customer loyalty through satisfaction and efficiency. Cogent Bus. Manag. **11**(1) (2024). https://doi.org/10.1080/23311975.2024.2326107
42. Tavares, D.R., Canciglieri Junior, O., Guimarães, L.B.M., Rudek, M.: A systematic literature review of consumers' cognitive-affective needs in product design from 1999 to 2019. Front. Neuroergonom. **1**(February) (2020). https://doi.org/10.3389/fnrgo.2020.617799
43. Varga, M., Albuquerque, P.: The impact of negative reviews on online search and purchase decisions. J. Mark. Res. (2023). https://doi.org/10.1177/00222437231190874
44. Vergura, D.T., Zerbini, C., Luceri, B., Palladino, R.: Investigating sustainable consumption behaviors: a bibliometric analysis. Br. Food J. **125**(13), 253–276 (2023). https://doi.org/10.1108/BFJ-06-2022-0491
45. Vizoso, Á., Vaz-álvarez, M., López-García, X.: Fighting deepfakes: media and internet giants' converging and diverging strategies against hi-tech misinformation. Media Commun. **9**(1), 291–300 (2021). https://doi.org/10.17645/MAC.V9I1.3494
46. Wan, J., Ma, H., Zhou, W., Qin, M., Li, P.: The study of female college students' consumer psychology mechanism toward male celebrity endorsed products: tempted or coerced? Heliyon **10**(9), e30401 (2024). https://doi.org/10.1016/j.heliyon.2024.e30401
47. Wang, L., Che, G., Hu, J., Chen, L.: Online review helpfulness and information overload: the roles of text, image, and video elements. J. Theor. Appl. Electron. Commer. Res. **19**(2), 1243–1266 (2024). https://doi.org/10.3390/jtaer19020064
48. Westerlund, M.: The emergence of deepfake technology: a review. Technol. Innov. Manag. Rev. **9**(11), 39–52 (2019). https://doi.org/10.22215/TIMREVIEW/1282
49. Whittaker, L., Letheren, K., Mulcahy, R.: The rise of deepfakes: a conceptual framework and research agenda for marketing. Australas. Mark. J. **29**(3), 204–214 (2021). https://doi.org/10.1177/1839334921999479
50. Whittaker, L., Mulcahy, R., Letheren, K., Kietzmann, J., Russell-Bennett, R.: Mapping the deepfake landscape for innovation: a multidisciplinary systematic review and future research agenda. Technovation **125**(November) (2023). https://doi.org/10.1016/j.technovation.2023.102784
51. Xiao, L., Li, Y.: Examining the effect of positive online reviews on consumers' decision making: the valence framework. J. Glob. Inf. Manag. **27**(3), 159–181 (2019). https://doi.org/10.4018/JGIM.2019070109
52. Yacout, O.M., Vitell, S.: Ethical consumer decision-making: the role of need for cognition and affective responses. Bus. Ethics **27**(2), 178–194 (2018). https://doi.org/10.1111/beer.12178

Deepfake Advancement: Navigating Technological Evolution, Ethical Dilemmas, and Regulatory Challenges

Harshi Garg, Mohammad Kashif, Priyank Sharma, and Abdullah Malik

Abstract Deepfakes have become an increasingly common phrase among judicial and IT Professionals. Deepfake and its evolution, ethical dilemmas, regulatory frameworks, and challenges are clarified in this book chapter. This study contributor investigates the deepfake advancement of one specific real-world case study on the corporate sector. The authors reveal that individuals and celebrities are frightened because of the harmful sides of deepfakes. The study uncovers community mindset within deepfakes, highlighting concerns regarding misuse and the urgent requirement for regulations. It accomplishes this through assessing major incidents, a real-world case study, and employing sentiment analysis using Natural Language Processing (NLP) tactics like VADER. The findings suggest that while deepfake can innovate, there are major moral issues connected with the advancement, needing a cautious approach to development and governance. This chapter recommends for preventive measures to mitigate the adverse effect of deepfake advancement on humanity and business, including stakeholder cooperation and public education.

Keywords Communications · Deepfake · Ethics · Regulatory framework · Technology

---

H. Garg
School of Commerce and Management, IIMT University, Meerut, India

M. Kashif (✉)
Graphic Era Deemed to Be University, Dehradun, India
e-mail: mkashif69@gmail.com

P. Sharma
IIMT University, Meerut, India

A. Malik
JAIN Online, Jain (Deemed-to-Be-University), Bengaluru, India

S. Taneja et al. (eds.), *Navigating the Deepfake Conundrum: A Manager's Roadmap*,
Information Systems Engineering and Management 44,
https://doi.org/10.1007/978-3-031-90742-5_12

1 Introduction

In the past few decades, deepfake technology has attracted and frightened individuals across the planet. The fakes are movement flicks or images modified to seem highly real utilizing powerful neural algorithms and technological computing. Although their appealing appearance has attracted focus, they also represent considerable psychological and economic threats [7, 14]. Counterfeits assess and analyze both audio and film details by applying artificial neural networks, and intense deep learning. We use enormous quantities of images, and natural collections for educational purposes, to recognize personalized patterns, attributes, and responses [16, 35]. Whenever this fake data appears, it generates extremely convincing and precise stuff. This presents severe moral issues regarding anonymity, solitude, and dignity.

There are dangers related to the capacity to successfully and efficiently modify and manipulate media materials, like the dissemination of inaccurate data and theft. Understanding the possible advantages and negative effects of deepfake data will become more crucial as it develops [4, 30]. With this information, people, organizations, groups, and governments may develop strategies to identify and prevent the negative effects of deepfake. The term Artificial intelligence, formerly referred to as advanced learning, stems from the expression of fake data [4].

Deepfake media intelligence exploits swapping faces in motion films, images, and other various kinds of online content to make deceptive items seem genuine. Algorithms powered by machine learning have been implemented in this way to show robots how to address difficulties through enormous quantities of information. To produce counterfeit data, two techniques were tested versus one another. Each is usually described as a sign, whilst the other is an instrument that generates authority [41].

1.1 The Evolution of Deepfake Technology

Experts who studied artificial intelligence computer vision in the mid-1990s have been associated with contributing to the development of the area of fraud. However, it was ineffective earlier in the 2010s, when General acceptance of significant deception was achieved. A significant inflection point in the field occurred in 2014 with the widespread availability of GANS and the continuing growth of computer-generated brains [38] (Fig. 1).

The classification tool has been tasked with determining whether the digital media material provided by the application indicates it is Artificial or genuine. When the examiner correctly recognizes whether the knowledge is authentic or bogus, it offers feedback to the creator to improve subsequent fake data. These methods converge to produce the generalized adversarial network (GAN) [39].

This trains one how to create associations through several methods, helping to accept the genuine traits needed for producing fake images. Deepfake is a form of

Fig. 1 Development of deepfake advancements. *Source* Authors'

online material whereby a single person's appearance is altered by the appearance of another individual in an image or picture. This similar member on Reddit first employed the word deepfake in the second quarter of 2017. This person developed an information site and aggregate device whereby they uploaded graphic images utilizing readily available face-swapping software. The phrase has now grown to include fake media products that appeared before the Reddit page and updated innovations [34].

The emergence of these GANs by individuals like good research fellows and others provided a foundation for fictional Data, which enabled ever-more complicated but convincing changes to be developed. In the beginning, deepfakes were solely employed by specialists and professionals due to the powerful processing capabilities and scientific expertise needed to create them. Older products, while interesting, lacked the reliability of false news [2].

As technology grew, fake information became more easily available. As open-source initiatives and the use of devices provide facilities, it becomes easier to develop convincing deepfakes. Although it is simple to find, deepfake continues to be developed and spread online at a rapid pace. Deepfakes that integrate both reality and fantasy have grown so realistic that they can sometimes be confused with genuine material [11].

The boost in absorption has been attributed to multiple advances, such as improved computer resources, enhancements in artificial intelligence techniques, and the enormous quantity of information provided for shaping these artificial neural networks. Deep learning-based systems that drive deepfakes have been extremely successful at recognizing and precisely replicating phrases, actions, and behavioral features. The capacity to change video and audio continuously constitutes one of the most significant and crucial advances in the discipline of deepfake [27].

These improvements have rendered live counterfeiting technologies possible, allowing users to adopt the real identity of someone else throughout online meetings or streaming. This pattern has profound implications that go outside mere pleasure and influence areas like political science, where worries related to possible power misuse are increasing [31].

The fakes developed as a consequence of the rising popularity of social networking sites and the increasing use of digital media. Since they amazingly make it accessible to millions of viewers to produce and appreciate videos, false products are an effective instrument for disseminating false data due to their wide availability and the propensity of individuals to trust anything they are subjected to [42].

1.2 The Dark Side of Deepfake Advancement: Deepfake Methods

The use of fake technology presents significant moral issues by amplifying fears regarding exploitation or misleading data. Deepfakes undermine conversation and confidence in the media by expertly combining counterfeit details with sincere ones. There are a few examples as Disinformation and alteration, Issues about vulnerability and security, Deception and personality fraud, and Implications on trust. Through the use of fake news, creators can generate impressively counterfeit videos or clip files, which contribute to disseminating inaccurate data to manipulate the public's view. As a consequence, it can lead to conflicts, damage others' credibility, and deceive individuals. Yet, many of those manipulated photos, clips, and visuals additionally represent an obstacle to the fraudulent right to free communication [19].

The capacity of technology to modify the way someone speaks or place a smile on a different individual is an important reason for worry. Individuals could unintentionally put themselves in undesirable or dangerous circumstances. In addition, an enormous amount, particularly photographs of licenses and other documents, can be generated, acquired, and preserved by companies and government departments and could be the focus of security thefts. Artificial intelligence is an important issue and can be highly damaging because it utilizes sensitive data that includes titles, confronts, or accents [28].

Table 1 Deepafake methods

Aspect	Description
Deepfakes in video	A multilayer network-based fake information automatic encoder analyzes the material of an original video to understand relevant features associated with the goal, including body language or the way they look. It subsequently patches the initial clip using these features. The encoding phase of this autonomous encoder hides all relevant characteristics, whereas the decoder is what inserts those features into the intended video
Deepfakes in audio	During sound deepfakes, a GAN replicates an audio capture, develops a representation using the speech trends, and executes the created template to generate any imagined speech sequence. Video game producers are frequently using this approach to create games
Syncing of lips	A different approach employed by counterfeits involves lip syncing g. In this instance, an audio file has been assigned to the footage through the deepfake, creating an appearance that the protagonist of the clip is saying the phrases displayed on the microphone. The visual element offers an additional level of deception assuming the sound is a deepfake. Ongoing network structures assist this kind of approach

Source Authors' compilation

The fakes may be exploited by a person having malicious goals to mimic other individuals, which might result in theft and stolen identities. Financial institutions, organizations, and people in general are not safe from scams or similar permanent repercussions like brand are monetary damage. The line between what is truly real as well as what is not may grow increasingly blurry as an outcome of such extremely realistic media offerings. Deep fakes have been identified in a variety of settings, including fraud, sociopolitical deceit, and theft [29] (Table 1).

1.3 Understanding of Deepfake Technology: Means of Communication Used to Generate Deep Fakes

Two encoder-decoder combinations are employed for this fake information output tactic. Throughout the purpose of training, two neural networks employ distinct decoding devices with utilize a single coder. To construct a deepfake, an image would be encoded using an average transmitter along translated via receiver. Deepfakes replicate human voices, facial expressions, and body language from real individuals through the use of an artificial brain [20] (Fig. 2).

A multitude or hundreds of thousands of pictures of the subject of interest are put into an algorithm throughout this procedure, and the information eventually gets encoded to alter or integrate the characteristics of a single motion picture with another. A GAN algorithm examines photos of the object being tracked through an assortment and viewpoints to gather all the data and opinions while creating a counterfeit picture [1].

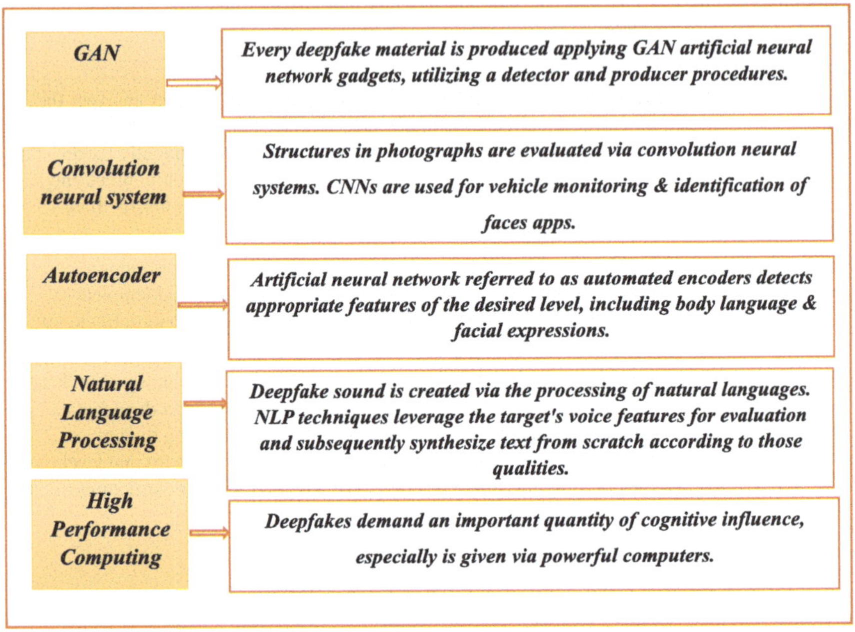

Fig. 2 Deep fake modes. *Source* Authors' compilation

The GAN technique scans the footage via numerous angles whilst producing a deepfake, besides analyzing movement, speech, and habits of behavior. To further enhance the ultimate photograph and footage's reality, this information is then sent via the algorithm for discrimination multiple times. Additionally, there are two distinct techniques employed to create deepfake videos. They can utilize the target's actual clip origin, whereby they can say and do stuff they have never done before, or they may execute an identity, where they put the target's profile onto a clip portraying somebody [1].

A variety of different technologies are frequently utilized for producing counterfeits in a matter of moments. According to the US Department report these tools are—Wav2Lip, Wombo, Zao, My Heritage, FaceApp, and FaceMagic [6].

1.4 Applications of Deepfake Technology

The deepfake technological applications remain proliferating steadily. Digitally produced counterfeit videos are commonly referred to as fake news. Through merging photographs, they may generate fresh data that reflects asserts, actions, or occurrences that transpired. The fakes are videos that have been created with algorithms for learning or digital applications. The outcome may additionally be extremely convincing. In addition, counterfeits are different from various other types of deception.

They are extremely difficult to come across. Material referred to as "deepfakes" usually take the appearance of speech, yet they may additionally adopt the form of vision. The process of deep learning serves to modify, adjust, or mix entities. As an outcome, it misleads spectators or listeners towards believing that something occurred or was said. Deepfakes are an improvement based upon the decades-old visual idea. A great deal of machinery, patience, and cinematic capacities are required for media mixing. On the other hand, audio fakes demand substantially less knowledge, dedication, and equipment—despite the reality that these fakes generally fail to stand out during the final examination. A few applications of deepfake technologies are as follows.

Entertainment Industry: Deepfake innovation has had an enormous effect on the field of entertainment. The directors offer numerous options because technological advances enable it feasible for animate realistic computer avatars. A notable example of such an act is the revival of musicians who have recently died seeking roles in upcoming films. There are some of the more prominent examples involving deepfakes within the field of entertainment. Considering the development of fake technological advancements, producers can now create films at lower prices while developing virtual images of renowned individuals and creatives who have gone. It not only cuts labor and time as well as allows followers to stay in touch with their favorites even after celebrities have passed on.

Acting Industry: Expert actors are currently being electronically duplicated for applications in motion pictures, and breakthroughs with deepfake Technology may boost the replicas' effectiveness or availability. Disneyland has used excellent quality deepfake swapping of faces tech to improve its animation effects. Disney upgraded its algorithm by including a face-swapping activity, employing gradual instruction intended to identify emotions, movement of the face, and repeating to maintain and enhance the outcome. Substantial production, as well as operational reductions, can be accomplished via this superior-resolution deepfake Technology.

Art Industry: The designer employed deepfake Technology to develop a computerized character who was a 1978–2018 time passenger, mimicking the 1980s film actress. This particular work of art was cited by the University of Massachusetts Institute during its collective intelligence analysis. The creator of the work delved into problems concerning the location of provoking within the creative sector and explored intergenerational perspectives via Ornella Muti's time voyage. Deepfakes are often employed within satirical and politicians along with celebrities' spoofs.

Memes and Dubbing Industry: A phony web trend that created pictures of actual individuals performing the chorus of "Baka Mitai" a tune of the music video, first came to light during 2020. Within the series, the participants deliver the melancholy tune throughout a cabaret minigame. Many of these viral memes derive their inspiration from a 2017 clip. Deep fakes deliver a cost-effective duplicating and translation option for the multinational television and film company. Using the capacity to sync performers' accents using different tongues, programming can potentially be consumed by a wider audience.

Education: Deepfakes are frequently applied by lecturers to render their teaching sessions more enjoyable. Likewise, such lessons would extend the boundaries of conventional media or formats that are visual. Intelligent computers may generate digital content which may bring historical life within the learning environment. Enhancing learning via involvement and communication. Employing a narration plus an opus depicting someone from history an imaginary documentary featuring replicas is going to work better. It could boost involvement and expedite the process of learning. Due to the small dimension and inexpensive price, artificial audio, and visual content may encourage pupils' advancement in school.

Cybersecurity and Politics: Deepfakes are categorized as attacks using social engineering within the field of cyber-crime because they possess a chance to expose private records and undermine the facilities of a company. Deepfakes are often employed to steal the identities of others through mimicking them. Several hundred thousand electors in Telangana, the state where votes were set to be held, got a clip on their cell phones which was divulged to them by a sitting minister appealing to them to cast their ballots for the prevailing government. There are several clips in the state of Madhya Pradesh which include sequences from the popular show Koan Bnega Crorepati. The short portrays the Amitabh Bachchan—hosted quiz program asking questions regarding politics in Madhya Pradesh to stir up spectators' dissatisfaction with rulers. The clips in both instances depict events that never occurred. The following clips are counterfeits that were created via artificial intelligence. The global community is about to enter an additional extremely dangerous period of deepfake politics, reflecting the expansion of fake news campaigns everywhere, and this triggered massive ballot divisions due to falsehoods.

2 Literature Review

2.1 The Gendered-Based Perspective

The gendered point of view revealed detailed neighbourhoods' fears regarding the damages that will be made to females and kids, which are the populations most at risk. This technique developed an argument that was basically concerning an absence of authority over how somebody behaves and their permission, to restrict false content to safeguard females from being bullied or harassed. There are multiple phonies, realistic, deepfakes showing prominent celebrities, like Rashmika Mandana, in apparently genuine pictures, with their real Visages artificially overlaid on the human figures of inappropriate material (Breitbart News, February 7, 2018). Precisely there has been a storm responding to signals suggesting that particular persons have produced illegal youngster maltreatment images using the pictures of under-16s, journalists identified the chance for misuse for anybody sharing their images on social networking sites (The Verge, February 9, 2018). Their continued acceptance of racist or gender-based stereotypes that are unresponsive (CNN, February 8, 2018) and their widespread fear of remorseful women celebrities and youngster renowned person (BBC News, February 7, 2018). The mainstream press informed audiences that "what's happening particularly troubling is the fact that a substantial prevalence of female celebrities who immediately achieved fame as teenagers regardless of a clear absence of recognition" (The Daily Telegraph, February 14, 2018).

Numerous people are currently devoting over 50% of their daily lives to the world wide web, currently a lot of females, the globe of advances can be a harmful setup the researchers mentioned on February 2, 2018. Networking site's decisions regarding limiting fake information should be regarded as a light of confidence in a generally grim narrative of online company inertia, specifically considering that counterfeits remain unaffected by judicial laws, stated the writers. It makes the internet more accessible place for females, which cannot be beneficial for everyone. As a consequence, an immense number of articles address the problem of sensual fake information on online platforms, pointing out the restriction as a means to strengthen community control and oversight. Realizing the effect that the media can have on the distribution of laws, we need to examine additional ways that media communicates policy for the internet throughout its entirety. It seems like journalists have indicated that the ban on counterfeits on the web will solve the matter of female abuse. This is a difficult approach due to communal fairness; technological products and their efficacy should not be regarded beyond the attached and recurrent myths they symbolize. Regulations covering unplanned habits, domestic assault, and gender disparities must be examined within the context of the ideological context of modern changes in interaction as an imminent risk.

2.2 *The Specialist's Perspective*

Regarding counterfeits, media groups produced a model perspective that covered the evaluation of the need for competent workers throughout a time of volatility in research, government, political discourse, and the creative industries. To comprehend this study, let's first analyze the concept of deepfake and fake news which has been frequently used by journalists to refer to the deepfake incidents. In the modern age of counterfeits and screening, VentureBeat (February 10, 2018) observed that these cases of spreading fake information contradict our common opinion regarding authenticity and may continue in establishing distance between widely agreed-up truths.

Furthermore, attracting the support of filters like Twitter will be the most efficient way to prevent this content from developing among individuals whose likenesses have been employed in the videos without their permission (The Verge, 7 February 2018). Considering the adverse reporting of deepfake information on web pages from this perspective, one may argue that presenting nearly terrible events helped to boost public trust in media research. It seemed to be that the press has protected itself from censure about the development of false innovation by putting the responsibility for the different levels of doubt and distrust that presently define civilization on the web, and also on people of every generation and elected officials. The basic and cultural differences between disclosure and cheating tactics were generated by this language.

2.3 The Realistic and Factual Perspective

This was unexpected to discover that in such a scenario, the traditional media designated counterfeits as a socially and politically destructive tendency. Authors offered counterfeit information in the type of footage (The Register, February 9, 2018) which were developed probably by AI systems that can identify the features of any person's face and connect it onto bodies in scenes (VICE, February 7, 2018) and are visible on the app store in just a few days, requesting pictures growth using neural networks that are artificial for general terms prospects entertainment" (Fox News, February 16, 2018). For instance, on February 21, 2018, a journalist claimed that AI advances are now able to relive realistic media content remotely. "They claimed that this innovation will effectively make it easy for deepfakes to spread untruth on the Internet. The progress raises issues on many levels, stated by TechCrunch writers, who added that the governed motion pictures reliance on machine learning systems will only become more advanced in the future years, which will make it harder for viewers to differentiate between actual and counterfeits content (TechCrunch, February 7, 2018).

The Verge's journalists on the subject replied, "if lip-sync and Face-swapping technologies are merged, an immense amount of deception is apparent. Based on The Verge (February 8, 2018), "this establishes a kind of digital indifference, an idea that no clip can ever be indisputably declared to be authentic." Thomas Jefferson once stated 'A well-informed population is a vital part of freedom'. The media at the news station stated, "How could we discern what is real?" while coming to their choices (February 20th, 2018). Nobody would completely believe what they heard in this particular instance, concluded the reporter (February 19, 2018).

The magazine identified fakes as a symbol of the dangers related to merging philosophical violent conflicts, automated economic customization, and profound intelligence. Deepfakes enable fresh methods to cover up fraud and tactical mistakes while harming actual indicators. Journalists accepted social media networks with counterfeits evidence in real-life situations, and this is exactly what we observed in gendered opinions. Fake news substance, nonetheless, produces a two-edged gadget for media, as we will discover in the paragraph that comes. Though fake news threatens society, they are a successful means for recovering trust among consumers in reporting.

2.4 Deepfake Videos Dissemination

The issue of how fraudulent data spreads is coming upwards, and seems that the internet has the response. Digital media supplies people's interest with an easy way to be updated, occupied, and socially involved. Untrue accusations arise out of nowhere. They generate a significant amount of curiosity on the web pages and are used by an extensive number of individuals. The credibility of many different nations and public organization bodies has been harmed by inaccurate reporting [37]. The primary reason for cyberbullying is that it undermines and eliminates powerful perspectives. Honest sharing of data and knowledge is challenged by copycat media, which also distorts legitimacy and genuine investigative variety. It can be observed to be a danger that could result in violence or death [36]. It can generate unhappiness and dissatisfaction caused by verdicts, which may ultimately affect highly educated individuals. Multiple videos are posted on the internet each day.

The inexpensive nature of websites and video editing tools has made it feasible to create and share deceptive movies with misleading materials. A segment that provides deceptive data and generates an unreliable film is regarded as an imitation clip. To detect variations that might indicate a lower degree of confidence in the accuracy of the pictures or indicate a desire for deepfake content, the present study suggests several kinds of examinations that require combining checks of a part of a video within the video itself [5]. False information, implies that it often comes along with concerns regarding the political consequences of an alteration in the way news is conveyed. Instagram and Twitter-related companies are receiving severe consequences for their part in propagating, assisting, and occasionally encouraging misleading content [24]. A few examples are news releases, edited images, overstated stories, the everyday show, and dishonesty. Certain groups produced YouTube videos that promoted generally accepted racist and discriminatory sentiments.

The particular data offered by openly accessible archives can also be suitable and improve civilization usually, but it may also regulate and impact the actions of individuals, which is required to promote major shifts in civilization [18]. These are the individuals who investigate and make changes to cultural growth, as shown by companion subcategories like these, which act as ideal examples for prospective SNS users of social networks. Similar information regarding SNS use and views of individuals and behaviors were to be collected [22]. It showed that clients preferred various places for relationships with those who had viewpoints comparable to their own. How can I evaluate whether or not the video footage is genuine? This document includes a few ideas and techniques, like constantly searching for genuine facts and news accounts, searching for the US page on the website, and examining reliable sources [13].

Investigators narrowed components that affect behaviors, and this research mainly examined the effect of consumer advantages, like perceived worth and anticipated use, on SNS communication, and how particular user actions may be affected by other fellow members of the collective [12]. The research examined the connection between assistance programs and the social impacts of cell phone SNS, and also its effect on messaging applications, databases of information, and social media platforms. The article explored how trust in misleading news affects theory, employing a two-response method. Employing the mixed process principle, they revealed how the topic encourages individuals to perceive it as untruth by isolating observation and dialogue, showing how the identical film extends across various mediums with comparable information but misleading headlines [8].

To capture the curiosity of consumers broadcasters are mindful yet less natural, which influences the accuracy of the viewed message. The extensive information of fake online phenomena and misinformation creates issues [9]. The possible impact of incorrect information on the finances, electoral politics, and social conditions highlights questions as to the reliability of discourse and decisions [3, 21]. Since false information relies on these basic values, preventing its negative impacts demands major action. Further study is required to examine the mental illness and digital parameters that contribute to the spread and adoption of untrue viral clips.

3 Research Method

The investigation method for the Facebook founder's Mark Zuckerberg fake deep-fake event provides a qualitative method to analyze the impact of Artificial media on business truthfulness, confidence in companies, and legislative problems. The research will employ a real-world case study to identify key trends that appear from different media sources reporting the loss of deepfake. Information will be collected using various publications, online media, feedback, remarks, and general discussion regarding the deepfake video.

Thorough analysis will be employed to group qualitative information into major categories which will be examined to understand the larger effects of deepfake advances. Sentiment analysis will also be carried out to assess user attitudes toward Facebook and deepfake Technology, employing Natural language processing (NLP) to evaluate remarks and posts across various websites. Case study and sentiment analysis will provide information on the incident's effect on the company image, the ethical difficulties, and regulatory solutions posed by deepfake advances (Fig. 3).

3.1 Research Objectives

1. To examine the general population's perception of Mark Zuckerberg's Deepfake case and how it influences general people's trust.
2. To analyze the ethical consequences of Deepfake technology and the requirement for regulatory structures that deal with challenges caused by content produced by Artificial Intelligence on social networks.

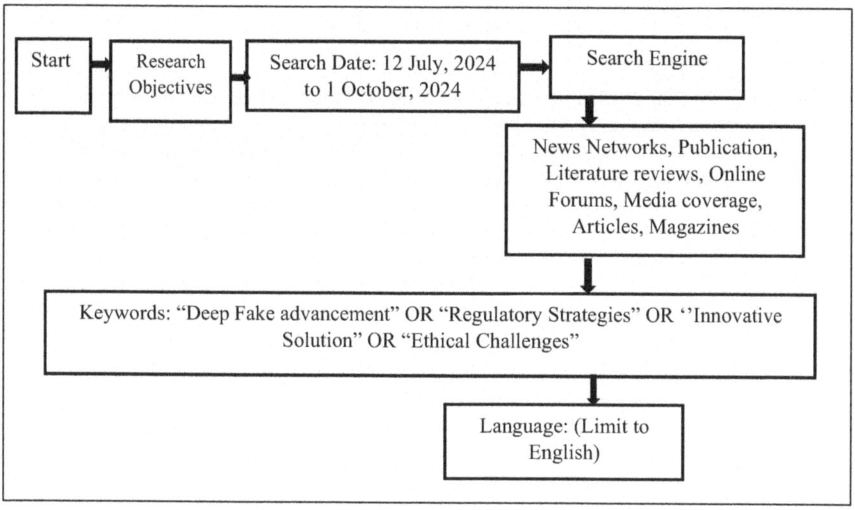

Fig. 3 Data extraction model. *Source* Authors'

3.2 Research Questions

1. How did the deepfake video of Mark Zuckerberg affect public belief in Facebook and social media platforms?
2. What Types of Ethical dilemmas are associated with deepfake technology and what Regulatory measures are necessary to combat the deepfake harm?

4 Data Analysis and Discussion

To accurately investigate the importance and response to the Mark Zuckerberg deepfake case, an ample number of second-hand sources of details have been reviewed as an element of the subject matter for the present investigation. The review was based on evidence obtained from news sources, magazines, academic evaluations, online communities, and expert remarks on social media. It includes qualitative and quantitative data (Table 2).

Media Coverage: The public service announcement received broad coverage in the press, including 100 pieces emerging on different websites during its initial month of broadcast. Prominent media outlets including 'The New York Times', 'Wired' and CNN contributed to raising its exposure. In the USA, major media outlets 'ABC News', 'CBS News' reported the PSA extensively. Extensive media coverage allowed the PSA to connect with a diverse worldwide readership and enhance its impact (Tables 3 and 4).

Table 2 Case study: Mark Zuckerberg deepfake incident

Attribute	Description
Full overview	In June 2019, a fake video of Facebook CEO Mark Zuckerberg was circulated on Instagram, portraying him as offering scary words regarding managing billions of individuals' information and having incomprehensible power. The clip was created by performers Bill Posters and Daniel Howe, in collaboration with the marketing organization Canny, to attack the company's privacy rules and promote attention concerning the possibility of misuse of deepfakes. This invented film used powerful artificial intelligence to change pictures and music, creating the appearance that Zuckerberg personally was talking about. The objective was to highlight the moral issues and legal difficulties posed by fakes, especially concerning deception and accountability for shareholders
Severity	Strong-however, no immediate or monetary damage occurred, and the incident created significant worries regarding corporate safety, management of reputation, and the likelihood of deepfakes spreading falsehoods with serious consequences
Harm type	The challenge to business ethics and public confidence—the deepfake information can undermine confidence in Facebook's capacity to regulate its website and avoid misinformation. It could negatively impact the business's credibility and image
AI system description	The deepfake video was produced with an adaptive adversarial network (GAN) alongside other artificial intelligence approaches. It employed influential neural systems to pair Mark Zuckerberg's facial characteristics with artificial audio, precisely recreating his speech and gestures. The method requires calls for practicing using deep learning techniques on hours of Zuckerberg clips to replicate his speech, tone, and gestures. The clip Alteration system 'Deep Face Lab' and the AI neural network 'GAN' were certainly applied to achieve the degree of accuracy displayed in the film. The method was sufficiently advanced to produce a possible fake, which deceived numerous viewers
System developer	The GAN system likely employed already trained samples from Zuckerberg's public presentations and remarks
AI functions	Artificial content creation—to create convincing fake information, the AI-generated fictitious audio and visual parts
AI applications	Misinformation and deception
AI tool	Generative Adversarial network (GAN)
Sector	Information and communication
Name	Mark Zukerberg deepfake video
Location	Instagram, social media site
Live lost	No
Physical system	Software only

(continued)

Table 2 (continued)

Attribute	Description
Data inputs	A great deal of freely available clip footage of Mark Zuckerberg, such as his remarks, discussion, and public speeches is likely what the deepfake was instructed on to generate a realistic depiction of his tone of voice and gestures
Level of autonomy	High

Source Authors'

Table 3 Media coverage report

Region	Main media outlets	Platform	Type	Impact
North America	The New York Times	Online, print, broadcast	News, tech	High (millions of readers)
	CNN			
	Wired			
	NBC News			
	ABC News			
	CBS News			
	Washington Post			
Europe	BBC	Online, broadcast, print	Current affairs, news	Moderate
	The Guardian			
	The Independent			
Asia	CNN Asia	Online, broadcast, print	News, regional focus	Moderate
	BBC Asia			
	South China Morning Post			
	The Japan Times			
Other regions	Al Jazeera	Online, broadcast, print	International news	High (worldwide audience)
	The Sydney Morning Herald			
	The Globe and Mail (Canada)			

Source Authors'

RQ1—Sentiment analysis was employed to examine general opinions of the ethical issues and legal proceedings related to deepfake advances. The present research aimed to investigate how people interpret the possible risks and advantages of fakes, especially in terms of deception and corporate transparency. To guarantee the

Table 4 Sentiment analysis

Source	Sentiment	Category	Overall sentiment scores
Twitter	Deepfakes pose a huge risk to individual privacy	Negative	−0.80
	Regulations might smother inventiveness in content creation	Negative	−0.50
Facebook	Regulating counterfeits can assist secure innocent individuals	Positive	+0.70
	The future of deepfakes to affect elections is alarming	Negative	−0.75
Online article (TechCrunch)	The advertising around deep fake is extravagant; they are not that challenging	Neutral	0.00
Reddit	There should be rules against deepfakes	Positive	+0.75
	Credibility is key; firms must reveal deepfake use	Positive	+0.70
Instagram	Many individuals are not knowledgeable of the dangers associated with deepfakes	Negative	−0.70
	I trust that deepfake can be employed without harm	Neutral	0.10

Source Authors'

outcome's reliability, a methodical approach was adopted, utilizing Natural Language Processing (NLP) methods and instruments like VADER (valence-aware dictionary and Sentiment Reasoner) or TextBlob. The initial stage includes acquiring qualitative information from multiple internet platforms, like social media responses, views, comments, posts, blogs, articles, and discussion boards, where individuals address deepfakes. The information was then prepared to eliminate unused data, address grammatical issues, and standardize the content for consistent analysis. Next, the cleared data was put into the VADER sentiment analysis application.

VADER is especially effective for sentiment analysis because it is designed to identify sentiment in online forums and brief texts while taking into account situational language particulars like slang, emojis, Grammar, and syntax. Once the information was examined by VADER, the software provided sentiment ratings to every remark, classifying them as positive, negative, and neutral. The measurement allowed a fair evaluation of the society's mood, revealing current beliefs and highlighting the necessity for regulatory regulations. Likewise, by summing up the sentiment values, we may figure out the aggregate pattern in public opinion. Sentiment analysis is a crucial technique in recognizing the greater ethical impacts of deepfakes, enabling knowledgeable discussions regarding suitable laws to avoid possible harm whilst promoting innovative endeavors.

RQ2—The examination was required for multiple reasons. First, it provided insight into the moral issues concerning fake advancement, showing the level of public concern and its misuse. The systematic approach ensured that the findings depended on actual information instead of subjective or created claims. Likewise, the inquiry highlighted the complex nature of public opinions, as numerous individuals expressed both fear and enthusiasm for novel applications of fake technology. This discrepancy indicates the value of bright regulations laws which maintain an equilibrium between innovation and values. The tables below underline numerous capabilities and limitations connected with deepfake approaches, which vary from scientific provocation to broad societal and ethical concerns, and provide some regulatory techniques for deepfake technology. Using numerous literature reviews regarding this technology and through media reports, published articles, and secondary data information authors have addressed the mitigation techniques of deepfake advancements.

5 Ethical Challenges of Deepfake Technology

The number of counterfeits on the World Wide Web increases every six months, demonstrating the rapid increase of fake material over the past few years. Roughly fifty thousand expression and image fakes have been shared on social networks globally in 2023, based on Deep Media. Considering an increase in counterfeits every six months, eight million deepfakes should be distributed worldwide by 2025. Deepfakes have grown prevalent due to the simplicity of getting hold of strong AI Tools and an abundance of data that is publicly accessible [26].

Deepfakes are computational intelligence and advanced training algorithm-generated simulated media outlets, often displayed as pictures, sound, or videos. These computer programs smoothly overlay a resemblance of one person over someone through modifying or altering current data leveraging massive database [15, 17]. The limited edition was primarily inspired by a 2018 trip to the state of Washington. In their now-famous Obama film, Seitz and his coworkers just released techniques that enabled just a few hours of fundamental sound recordings to generate a video with excellent lip-syn.

An animated portrayal of Obama, double his age, perfectly replicates the young Obama sound recording at the final moment of the video. It's likely perhaps the most iconic deepfake footage, even if it's not the one that started it all. It is rendered feasible by the use of machine learning, which converts language onto "mouth forms" by employing an array of films as a set for learning. These forms are subsequently integrated into an already-existing subject clip [40].

The multiple psychological and social issues that such counterfeits brought up, including if watching them destroys trust in the press, were another eye-opener. In what manner could fake news be used in the context of society? Exist techniques for

Table 5 Ethical challenges

Aspects	Description
Exploitation of society	Significant environments are currently being disrupted by counterfeits that depict significant individuals expressing themselves or interacting in ways that actually will. A famous example of this happens when Volodymyr Zelenskiy, who is also the nation of Ukraine, urges his soldiers to cease shooting in a fake clip
Bogus data and media reports	Disinformation has become an important issue, and it has become exacerbated by the rapid spread of counterfeits on the web. Based on new studies by the College of Chicago along with the information security firm, CHEQ, misleading information will ultimately cost the entire globe's commerce $78 billion by 2020
Privacy issues	Deepfakes possess a chance to prove detrimental whenever used for activities such as creating bogus material involving persons with no authorization from them. A significant example of such involves the 2023 actors' walkout, whereby actresses revolted from the unauthorized misuse of their physical by AI and fake news
Influence on organization	Deepfakes have a lasting effect irrespective of the world of business. Companies are confronted with further barriers to maintaining their goodwill and establishing faith, such as CEOs for frauds to the generation of counterfeit content

Source Authors' compilation

disputing or countering fakes? Social researchers have taken a look at the social implications of the technological advances at this point, notwithstanding the reality that computational researchers have performed a lot of work on automated manufacturing or identification of deepfakes [25].

It is time to understand how the field of psychology and media concepts pertain to deepfakes and what possible repercussions they could have on people. Deepfakes offer significant challenges for numerous areas of the world, especially government and business. Listed below are a few important ways in which counterfeits are currently influencing our daily lives (Tables 5, 6 and 7).

6 Regulatory Compliance, and Legal Frameworks

On the ninth of December, 2023 the European Union parliament and the EU cabinet achieved a temporary compromise regarding anti-inflation legislation. This legal document provides a broad structure that encompasses four separate threat divisions: bare or low risk, bound threat, higher danger, and inappropriate harm. Within these discovered risk areas, the AI policy desires to fully address the possible threats presented by artificial learning and AI. Regardless of a few certain exceptions for authorized workers, such as rejection, the AI software would employ a risk-based

Table 6 Advantages of deepfake technologies

Advantages	Description
Construct pragmatic videos and description	Deepfake innovation can produce extremely persuasive videos and depictions of a person plus adjusting facial appearance, diction, and gesture
Create artificial aural	A deepfake system can produce nonnatural sound that replicates an individual's sound, validating the design of pragmatic commentary and style
Promote facial agitator	The deepfake approach permits the depicting of an individual's facial looks and gestures onto different individuals' faces in clips, validating practical imitation
Amplify exceptional sequel in films	A deepfake system can be employed to increase optic sequels in films, producing faithful CGI personalities or changing clips with virtual manipulation
Evaluate computer graphics tasks	The deepfake approach can aerodynamic computer graphics procedure by mechanical functions like lip-syncing or facial vibrancy
Authorize effective personalities and images	Counterfeit processes can generate realistic digital personalities and images for video games, digital reality, and other implementations
Support in facial energy	Deepfake systems can be employed for facial expression and depiction, permitting truthful gestures of digital Images and avatar figures

Source Authors' compilation

Table 7 Limitations of current deepfake technologies

Limitations	Descriptions
Discernment provocation	Deepfake clips can be hard to catch employing conventional practice, requesting modern procedures and apps for recognition
Moral concern	The deepfake system elevates important moral concerns, plus prospective misuse for fraud, persecution, or increasing delegation
Misuse of spiteful aims	Deepfakes can be utilized for malevolent motives like producing gossip, information, publicity, advocacy, and brainwashing
Legal entanglement	The service of deepfake strategy may increase challenges concerning safety, patent, trademark, and aspersions laws
Attribute limitation	The attributes of deepfake videos and pictures can differ, with few examples manifesting antiquities, unpredictability, or abnormal motion
Figuring concern	Counterfeit cohort usually needs notable reckoning expedient, restricting convenience for peoples or firms with finite graphic power
Future of social exploit	Deepfakes have the future to change general opinion, threaten belief in media, and aggravate communal conflict

Source Authors' compilation

strategy, classifying imitates as horrible risk and applying moderate transparency obligations to show that the data was produced using analytics [32].

Legislative proceedings started in 2021, but innovations in machine learning like ChatGPT transformed the category with innovative features. Although these virtual brain networks can generate economic analysis strikes, there are further risks linked to such systems when looking at their dark sides. Fake innovation is one of those problems that get more troubling with every passing hour. The EU has decided not to expressly develop a regulatory structure for AI to deal with this particular fear. It is the globe's earliest literature of this kind [23].

The procedure took longer than intended since guidelines regulating innovation and its boundaries occasionally fall behind developments in advances. Nations like France and Germany, though, have genuine arrangements about these rules, believing that severe monitoring will hinder innovation and damage potential AI entrepreneurs. Mistal AI, a French start-up is creating AI algorithms, which is essential to these kinds of business as the global AI rivalry increases. While the recent policy is certainly significant and innovative-thinking, it is also essential. Possibly there will be other advances in AI rules until the law is formally adopted by the parliament and member countries [4, 10].

While deepfakes are not specifically banned by law in India, numerous restrictions may control deepfake at large to a certain level. Deepfakes are regulated somewhat by the seventh rule of the IT code, 2021, along with other specific IT Act laws. Article 79(1) shields web intermediaries from becoming responsible for separate information, material, or activities on their web pages. This system of justice achieves the suitable equilibrium by providing citizens with a remedy in law while protecting facilitators from private responsibility for players' details.

A device fraud and abuse section indicate the implications of sending or sharing images that are another person's private secret without the person's approval, which represents an economic breach. The punishment includes a financial penalty of INR 2 lakh rupees and 3 years in jail. The previously mentioned initiative protects individuals against online security risks by looking to discourage and avoid the illegal publication of critical imagery. The moral dilemma that is confronted by deep fakes, revolves around if this rule can be used when the ensuing images are completely forged and developed employing creative Ais [33].

Internet fraud and abuse, 67, 67A, explicitly forbid and punish the online spread or distribution of inappropriate content, material showing bodily aggressive content, and information including youngsters participating in such conduct, etc. The goal of these rules and regulations is to avert the circulation of damaging and incorrect substances by guaranteeing that those who do so will be punished. A new legislation was implemented, illustrating the current administration's inadequate reaction to the continuing issues. It establishes the requirements of information consultants, especially seeking legitimacy from a data user before processing any of the information.

7 Deepfake Detection and Mitigation Technique

Based on an Attesiv poll, 82% of those polled think that counterfeits might represent a danger for their business, while more than 46% believe that their business doesn't have an approach established for dealing with this risk. As counterfeits get increasingly unsafe, business leaders and cybersecurity professionals have to come together to develop an extensive strategy for protecting the businesses they manage. Preventive safeguarding shouldn't be postponed since fake information assaults may impact any sort of organization. A few suggestions and guidelines may assist your safeguarding on such hazards.

7.1 Maintain Effective Cybersecurity Administration and Employee Training

Deepfakes have adverse impacts on business and can be substantially mitigated through efficient security management. To prevent this risk, an extensive set of rules and processes is necessary. Following all relevant legislation on data security and confidentiality is an essential aspect. A quarterly inspection of security is helpful to identify whether adjustments may be made or not. While counterfeits are technological goods, human beings remain essential for them to be effective.

Consequently, businesses must educate their workers on the hazards of counterfeits and how to recognize such fake content. Identifying deepfakes needs to be one of the topics addressed in technological education programs. Creating an office environment in which individuals verify and scrutinize things is important. Businesses ought to emphasize that employees have to investigate sources of knowledge and verify the authenticity of media materials looking for electronic watermarks or inconsistencies in video or audio files.

7.2 Strengthen Identification Methods

Companies should reinforce their personal information authentication process as well as establish an alternate method for obtaining financial credit card clearance to deal with the increasing danger of deepfakes.

Companies ought to invest in advanced risk detection equipment with the capacity to identify deepfakes to safeguard themselves against the potential risk presented by such counterfeits. One must maintain a look for signs of fake news that might harm the business image. Especially when it comes to digital content. It is essential to act immediately on potential erroneous or unsafe material to prevent any prospective repercussions.

7.3 Facial Recognition, Movement, and Audio Analysis

Modern technologies for facial recognition are employed to identify irregularities in facial features. Considering its value, facial scanning has multiple limits. Excellent workmanship may sometimes escape up-to-date identification computation.

This involves checking the Individual's measures and emotions for any signs of falsehood, such as unusual head motions or smiles that are inconsistent with the language being used. Fake news identification depends largely on sound analysis, which searches for deviations in speech tone, pronunciation, and lip-sync problems. This involves reviewing and verifying the Clip or image's content, origins, history, and other characteristics for deviations.

8 Conclusion

Deepfakes are showing equally the artistic and harmful options of artificial media. The dual function of deepfake highlights the value of knowing this advances, not only what it is and how it operates, but additionally its prospective implications. The continuous development of deepfakes requires evolving and active tactics. Constant attention is essential for developing and enhancing techniques for detection, rules and regulations, and moral standards. Comparably, we need to improve our digital understanding. By embracing both a sense of responsibility and duty of care, we can overcome the challenges created by deep fakes, assuring that we benefit from the opportunities they present while preserving our community's trust in the online realm. As we conclude our research into the global phenomena it's obvious that this type of technology presents an essential juncture in the online age.

The journey exploring the development, evolution, implementation, and probable futures of fake news reveals the complex network of problems and the chances they provide. Counterfeits are more than basically an academic miracle; they represent our society's view of reality, morality, and realism in the online era. Recognizing the complicated nature of fakes is crucial, not merely for their scientific ramifications, but also their bigger legal, societal, and private rights effects. The increasing number of counterfeits has highlighted the urgent need for an integrated plan that spans sectors. Fakes deliver an issue that technological companies, legislators, academics, and consumers have to collaborate to tackle.

The collaboration is essential to creating successful detection methods, establishing solid rules and regulations, and increasing public consciousness. Digital companies play a significant part in establishing detection mechanisms in their applications to promote responsible artificial intelligence. Politicians and legal scholars play a crucial part in creating regulations and legislation that combine entrepreneurship with security and confidentiality issues. Scientists and teachers maintain the flame of information, growing the comprehension of fake and educating people to find reality in the world of technology.

The importance of the public significance cannot be ignored. In a period when information is easily accessible, it is up to all of us to figure out and share the truth. Establishing a keen eye for online data, investigating sources, and keeping conscious regarding the nature of fakes are all essential steps that all individuals can take. Both consumers and producers of digital media, our behavior and understanding serve to establish an online community founded on openness and reliability.

References

1. Ahmed, S.: Who inadvertently shares deepfakes? Analyzing the role of political interest, cognitive ability, and social network size. Telematics Inform. **57** (2021). https://doi.org/10.1016/j.tele.2020.101508
2. Ahmed, S.: Navigating the maze: deepfakes, cognitive ability, and social media news skepticism. New Media Soc. **25**(5), 1108–1129 (2023). https://doi.org/10.1177/14614448211019198
3. Ahmed, S., Ng, S.W.T., Bee, A.W.T.: Understanding the role of fear of missing out and deficient self-regulation in sharing of deepfakes on social media: evidence from eight countries. Front. Psychol. **14** (2023). https://doi.org/10.3389/fpsyg.2023.1127507
4. Albahar, M., Almalki, J.: Deepfakes: threats and countermeasures systematic review. J. Theor. Appl. Inf. Technol. **97**, 22 (2019). www.jatit.org
5. Amerini, I., Barni, M., Battiato, S., Bestagini, P., Boato, G., Bonaventura, T.S., Bruni, V., Caldelli, R., De Natale, F., De Nicola, R., Guarnera, L., Mandelli, S., Marcialis, G.L., Micheletto, M., Montibeller, A., Orru', G., Ortis, A., Perazzo, P., Puglisi, G., Vitulano, D., et al.: Deepfake Media Forensics: State of the Art and Challenges Ahead (2024). http://arxiv.org/abs/2408.00388
6. Arslan, F.: Deepfake technology: a criminological literature review. Sakarya Üniversitesi Hukuk Fakültesi Dergisi **11**(1), 701–720 (2023). https://doi.org/10.56701/shd.1293642
7. Barari, S., Lucas, C., Munger, K., Huff, C., Lelkes, Y., Webster, S.: Political Deepfake Videos Misinform the Public, but No More than Other Fake Media (2021)
8. Bateman, J.: Deepfakes and Synthetic Media in the Financial System: Assessing Threat Scenarios Cyber Policy Initiative Working Paper Series | "Cybersecurity and the Financial System" #7 (2020)
9. Battista, D.: Political communication in the age of artificial intelligence: an overview of deepfakes and their implications. Soc. Regist. **8**(2), 7–24 (2024). https://doi.org/10.14746/sr.2024.8.2.01
10. Capistran, J.B.: Evolution of deepfake: semantic fields and discursive genres (2017–2021). Icono14 **20**(1) (2022). https://doi.org/10.7195/ri14.v20i1.1773
11. Debroy, O., Bhargavi, D., Hemmige, D.: Psycho-social impact of deepfake content in entertainment media. Mon. Peer-Reviewed Refereed Index. J. IC Value **86**, **87**(10), 2455-0620 (2024). https://doi.org/10.2015/IJIRMF/202405032
12. Ekpang, J.E., Iyorza, S., Ekpang, P.O.: Social media and artificial intelligence: perspectives on deepfakes' use in Nigeria's 2023 general elections. Kampala Int. Univ. Interdiscip. J. Hum. Soc. Sci. **4**(2), 109–124 (2023). https://doi.org/10.59568/KIJHUS-2023-4-2-08
13. Galyashina, E.I., Nikishin, V.D.: The protection of megascience projects from deepfake technologies threats: information law aspects. J. Phys.: Conf. Ser. **2210**(1) (2022). https://doi.org/10.1088/1742-6596/2210/1/012007
14. Gamage, D., Chen, J., Ghasiya, P., Sasahara, K.: Deepfakes and Society: What Lies Ahead?, pp. 3–43 (2022). https://doi.org/10.1007/978-981-19-1524-6_1
15. Garg, H., Ike, G.N., Kashif, M., Azam, M.S.: Dynamics of sustainable and ethically responsible business practices in corporate social responsibility. In: Technology-Driven Evolution of the

Corporate Social Responsibility Ecosystem, pp. 92–113. IGI Global (2024). https://doi.org/10.4018/979-8-3693-3238-2.ch005
16. George, A.S., Hovan George, A.S.: Deepfakes: The Evolution of Hyper Realistic Media Manipulation (2023). https://doi.org/10.5281/zenodo.10148558
17. Godulla, A., Hoffmann, C.P., Seibert, D.M.A.: Dealing with deepfakes—an interdisciplinary examination of the state of research and implications for communication studies. Stud. Commun. Media **10**(1), 73–96 (2021). https://doi.org/10.5771/2192-4007-2021-1-72
18. Hancock, J.T., Bailenson, J.N.: The social impact of deepfakes. Cyberpsychol. Behav. Soc. Netw. (Mary Ann Liebert Inc.) **24**(3), 149–152 (2021). https://doi.org/10.1089/cyber.2021.29208.jth
19. Hussain Al-Khazraji, S., Hadi Saleh, H., Ibrahim Khalıd, A., Adnan Mıshkhal, I.: Impact of deepfake technology on social media: detection, misinformation and societal implications. Technol. Sci. **23** (2023). www.isres.org
20. Ishtiaq, S., Azlina Abd Rahman, N., Shajaratuddur Harun, K.: Cyber security and threats: deepfakes impacts and risks. J. Appl. Technol. Innov. **6**(2) (2022)
21. Kashif, M., Kumar, P., Ghai, S., Kumar, S.: Disruptive technologies in computational finance. In: Algorithmic Approaches to Financial Technology: Forecasting, Trading, and Optimization, pp. 46–60 (2023)
22. Kashif, M., Shajar, S.N., Singhal, N., Kumar, P.: Achieving sustainable investment practices through green finance: challenges and opportunities. In: Sustainable Investments in Green Finance, pp. 234–244 (2023)
23. Kashif, M., Garg, H., Weqar, F., David, A.: Regulatory strategies and innovative solutions for deepfake technology. In: Navigating the World of Deepfake Technology, pp. 262–282. IGI Global (2024). https://doi.org/10.4018/979-8-3693-5298-4.ch013
24. Kwok, A.O.J., Koh, S.G.M.: Deepfake: a social construction of technology perspective. Curr. Iss. Tour. (Routledge) **24**(13), 1798–1802 (2021). https://doi.org/10.1080/13683500.2020.1738357
25. Lee, Y.A., Huang, K.T.T., Blom, R., Schriner, R., Ciccarelli, C.A.: To believe or not to believe: framing analysis of content and audience response of top 10 deepfake videos on YouTube. Cyberpsychol. Behav. Soc. Netw. **24**(3), 153–158 (2021). https://doi.org/10.1089/cyber.2020.0176
26. Leibowicz, C., Mcgregor, S., Ovadya, A.: The Deepfake Detection Dilemma: A Multistakeholder Exploration of Adversarial Dynamics in Synthetic Media the Partnership on AI 2 XPRIZE Foundation, Syntiant Corp 3 Thoughtful Technology Project (2021)
27. Mahashreshty Vishweshwar, S., Vishweshwar, M.: Implications of Deepfake Technology on Individual Privacy and Security Recommended Citation (2023). https://repository.stcloudstate.edu/msia_etds
28. Manuel, G.I., Sá, V.J., Gomes, P.V., Marques, A., Pereira, J.: Study of deepfakes in cyberspace, impact, and consequences on the social environment. Atas Da Conferencia Da Associacao Portuguesa de Sistemas de Informacao 221–233 (2022). https://doi.org/10.18803/capsi.v22.221-233
29. Nagumotu, K.: Deepfakes Are Taking over Social Media: Can the Law Keep Up?, vol. 62 (2018)
30. Neethirajan, S.: Is seeing still believing? Leveraging deepfake technology for livestock farming. Front. Vet. Sci. (Frontiers Media S.A.) **8** (2021). https://doi.org/10.3389/fvets.2021.740253
31. Nowroozi, E., Seyedshoari, S., Mohammadi, M., Jolfaei, A.: Impact of media forensics and deepfake in society. In: Breakthroughs in Digital Biometrics and Forensics, pp. 387–410. Springer International Publishing (2022). https://doi.org/10.1007/978-3-031-10706-1_18
32. Qureshi, S.M., Saeed, A., Almotiri, S.H., Ahmad, F., Ghamdi, M.A.A.: Deepfake forensics: a survey of digital forensic methods for multimodal deepfake identification on social media. PeerJ Comput. Sci. **10**, 1–40 (2024). https://doi.org/10.7717/PEERJ-CS.2037
33. Ramachandran, V., Hardebolle, C., Kotluk, N., Ebrahimi, T., Riedl, R., Jermann, P., Tormey, R.: A multimodal measurement of the impact of deepfakes on the ethical reasoning and affective reactions of students. In: SEFI 2023—51st Annual Conference of the European Society for

Engineering Education: Engineering Education for Sustainability, Proceedings, pp. 1122–1132 (2023). https://doi.org/10.21427/EAJR-WE79

34. Samuel-Okon, A.D., Akinola, O.I., Olaniyi, O.O., Olateju, O.O., Ajayi, S.A.: Assessing the effectiveness of network security tools in mitigating the impact of deepfakes AI on public trust in media. Arch. Curr. Res. Int. **24**(6), 355–375 (2024). https://doi.org/10.9734/acri/2024/v24i6794
35. Shahzad, H.F., Rustam, F., Flores, E.S., Luís Vidal Mazón, J., de la Torre Diez, I., Ashraf, I.: A review of image processing techniques for deepfakes. Sensors **22**(12). MDPI (2022). https://doi.org/10.3390/s22124556
36. Shajar, S.N., Kashif, M., George, J., Nasir, S.: The future of green finance: artificial intelligence-enabled solutions for a more sustainable world. In: Harnessing Blockchain-Digital Twin Fusion for Sustainable Investments (2024). https://doi.org/10.4018/9798369318782.ch013
37. Taylor, B.C.: Defending the state from digital deceit: the reflexive securitization of deepfake. Crit. Stud. Media Commun. **38**(1), 1–17 (2021). https://doi.org/10.1080/15295036.2020.1833058
38. Temir Asst Kastamonu, E.: Deepfake: new era in the age of disinformation & end of reliable journalism. Selçuk İletişim **13**(2), 1009–1024 (2020). https://doi.org/10.18094/JOSC.685338
39. Vasist, P.N., Krishnan, S.: Engaging with deepfakes: a meta-synthesis from the perspective of social shaping of technology theory. Internet Res. **33**(5), 1670–1726 (2023). https://doi.org/10.1108/INTR-06-2022-0465
40. Whittaker, L., Letheren, K., Mulcahy, R.: The rise of deepfakes: a conceptual framework and research agenda for marketing. Australas. Mark. J. **29**(3), 204–214 (2021). https://doi.org/10.1177/1839334921999479
41. Whittaker, L., Mulcahy, R., Letheren, K., Kietzmann, J., Russell-Bennett, R.: Mapping the deepfake landscape for innovation: a multidisciplinary systematic review and future research agenda. Technovation **125** (2023). https://doi.org/10.1016/j.technovation.2023.102784
42. Wu, F., Ma, Y., Zhang, Z.: I found a more attractive deepfaked self: the self-enhancement effect in deepfake video exposure. Cyberpsychol. Behav. Soc. Netw. **24**(3), 173–181 (2021). https://doi.org/10.1089/cyber.2020.0173

Dynamics of Deep Fake in Higher Education

Rajesh Tiwari, Bimal Anjum, Khem Chand, and Shweta Bajaj

Abstract Deep Fake has attracted the attention of stakeholders. Higher Education is not isolated from the disruption caused due to deep fake. Young minds are eager to experiment with new technology for learning and completion of academic assignments. The Chapter reviews the dynamics of deep fake in higher education. The published literature was reviewed to examine the issues underlying the deep fake in higher education. Google trends were evaluated to explore the interest of the stakeholders. It was found that deep fake is a double-edged sword for the higher education. It enhances the efficiency of the students and provides opportunity to explore multiple issues using deep fake technology. On the other hand, it deep fake has serious ethical issues along with plagiarism. The teachers need to be trained so that they can mentor students on judicious usage of deep fake technologies.

Keywords Deep fake · Higher education · Plagiarism · Students · Efficiency · Teachers

R. Tiwari (✉)
Department of Management Studies, Graphic Era (Deemed to Be University), Dehradun, Uttarakhand, Bharat, India
e-mail: ambitioncfarajesh@gmail.com

B. Anjum
DAV College, Chandigarh, Bharat, India

K. Chand
MM Institute of Management, Maharishi Markandeshwar (Deemed To Be University), Ambala, Bharat, India

S. Bajaj
Professional Communication Department, Graphic Era (Deemed to Be University), Dehradun, Bharat, India
e-mail: shwetabajaj.comm@geu.ac.in

S. Taneja et al. (eds.), *Navigating the Deepfake Conundrum: A Manager's Roadmap*, Information Systems Engineering and Management 44,
https://doi.org/10.1007/978-3-031-90742-5_13

1 Introduction

The deep fake has attracted attention of various stakeholders since it gained popularity in 2017, through swapping of faces using artificial intelligence by Reddit user. Development of Generative Adversarial Networks by Goodfellow and his colleagues was a major breakthrough. Deep fake is comprised of deep learning and fake. The film Rogue One: A Star Wars Story (2016) was instrumental in highlighting the significance of deep fake. Now there is widespread use of deepfake.

Technology is a double-edged sword for students; it can have adverse effects of plagiarism and positive effects of improving efficiency [14]. Technology is getting integrated in the learner production outcomes [6]. Various studies have attempted to explore the role of AI in education [9]. Curriculum does not yield long term improvements in critical thinking among college graduates [12]. It is difficult but possible to check the adverse impacts of deep fake [10]. Latest technologies including AI has gained attention of the governments [22]. Students are not able to differentiate between original and deep fake videos, in spite of prior understanding that they have been shown deep and original videos [7]. Educators need to be made aware and trained to identify deep fake so that they can sensitise the student community [2]. Deep fake can be used to develop immersive educational videos [11]. Instead of banning technology tools for students, they should encourage to use it properly to leverage it to prepare for future challenges [19]. Generative AI has led to ethical issues and production of low quality of research output [3].

Willingness to use deep fake without proper understanding can have adverse implications [8]. Fusion of deep fake and traditional teaching pedagogies need to be explored [1]. The curriculum needs to integrate deep fake technology to prepare the students to face the challenges of the future [20]. Deep fake has significant impact on diverse stakeholders [18]. Deep fake has both merits and demerits. The long-term outcome for the society will depend on the understanding of ethical and governance issues [15]. Though students consider ethical issues as significant issues of deep fake and large language models, students prefer to use them for the benefits in academic pursuits [23]. The availability of larger sets of databases will lead to development of improved features in deep fake technologies [16].

2 Trend of Deep Fake Search on Search Engine

The Google trend results show that interest in deep fake has increased significantly in recent years. The Google trend of search for deepfake is shown in Fig. 1. The trend has grown significantly from October 2017.

The search trend for deep fake in India has increased significantly since 2018 and was at peak in November 2023 as shown in Fig. 2. There are reginal differences in the search of deep fake in India. The search is led by North East and Southern states as depicted in Fig. 3. Northern states show lowest search figures. The Hindi speaking

Fig. 1 World Google search trend of deep fake. *Source* Google Trends

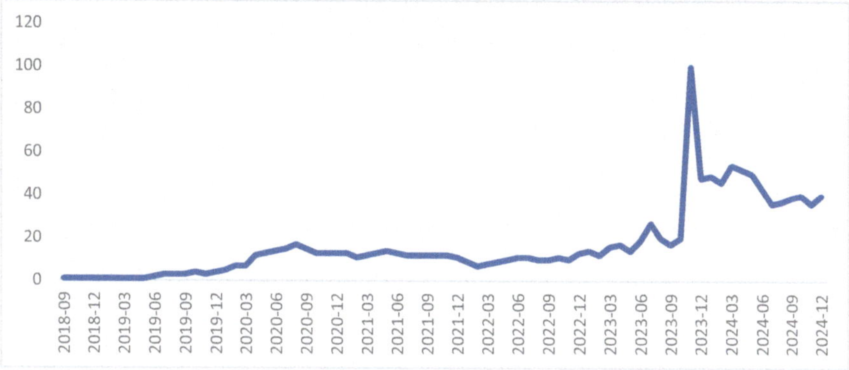

Fig. 2 Google search trend of deep fake in India (2004–2024). *Source* Google Trends

states of Madhya Pradesh, Uttar Pradesh, Bihar, Rajasthan total search (66) is even less than Mizoram (100). The regional differences need to be examined with regard to national security matters and cyber frauds.

The trend of Google search for deep fake videos is shown in Fig. 4. There is gradual rise in the interest in deep fake videos since 2018, but there is sharp increase in 2023.

Similar trends are witnessed in India. The interest in deep fake videos has shown significant increase since 2018. The rise is sharper in India as compared to the rest of the world as depicted in Fig. 5.

Though the interest for deep fake learning picked up in 2018 in world, in India it was observed in 2020 as shown in Figs. 6 and 7.

Lack of legal framework for deep fake is a cause of concern [17]. Cyber security issues for students need to be given due consideration [13]. Deep fake in wrong hands can be dangerous for the law and order in society [5]. Deep fake has immense possibilities in diverse field such as education, medical, entertainment. Constructive use of the technology underlying deep fake algorithms need to be explored for reducing cost, enhancing accessibility and making this world a better place [21]. The positive aspects of deep fake technology have been ignored [4].

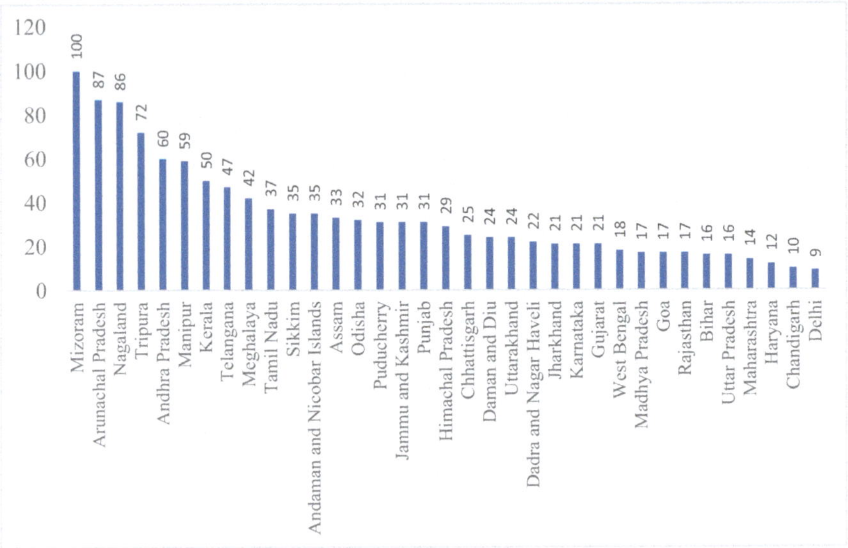

Fig. 3 State wise Google search trend of deep fake in India (2004–2024)

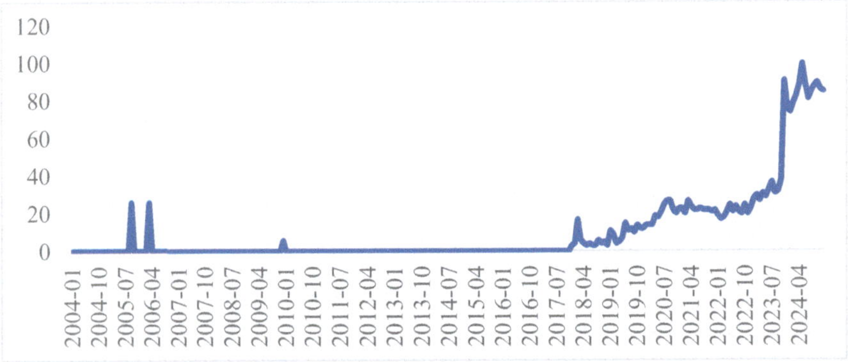

Fig. 4 World Google search trend of deep fake videos (2004–2024)

3 Impact of Deep Fake for Critical Thinking

Deepfakes, manipulated digital content using artificial intelligence in producing highly realistic but false images, videos, or audio, have sparked interest from diverse groups including in the field of education. The profound effects that deepfakes could impose on students' focus and their critical thinking skills have become a serious concern of observation. This review, therefore, discusses the positive impact and negative effects that deepfakes have imposed on the cognitive aspects of the students and their capacity for learning.

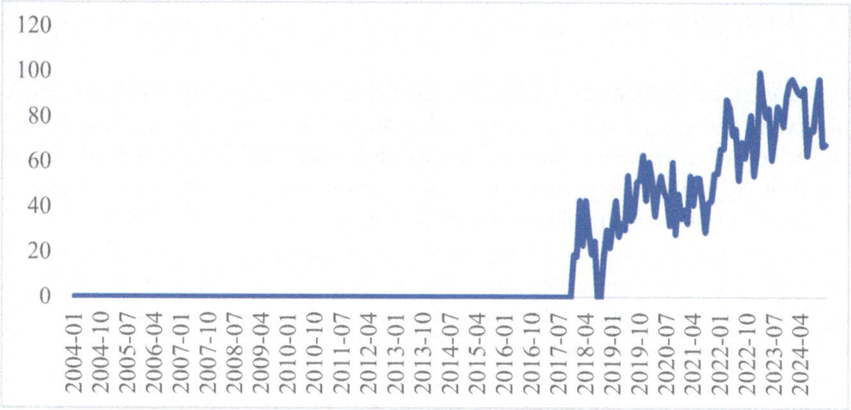

Fig. 5 Google search trend on deep fake videos in India (2004–2024)

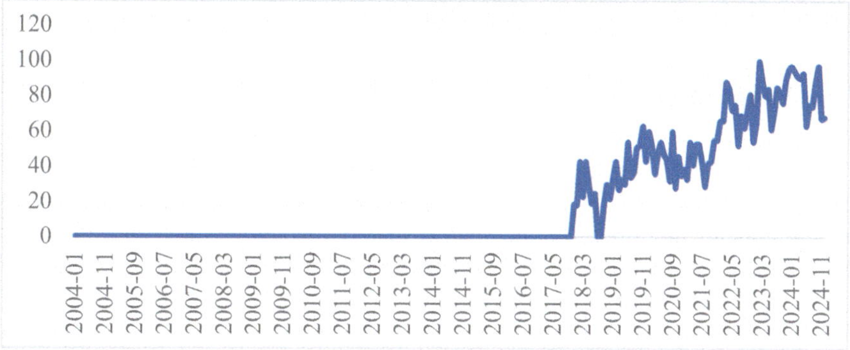

Fig. 6 Google search trend on deep fake learning in the world (2004–2024)

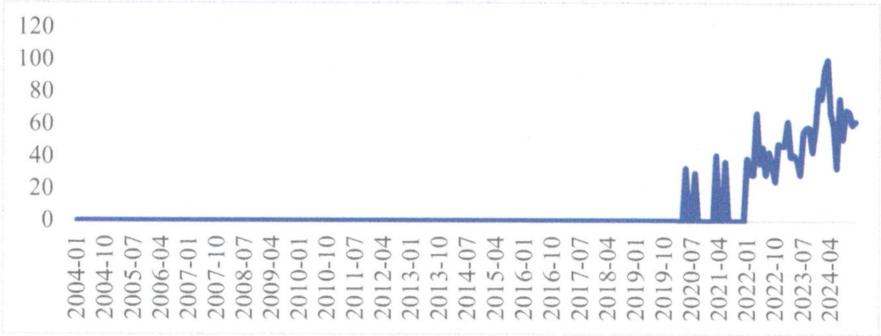

Fig. 7 Google search trend on deep fake learning in India (2004–2024)

3.1 Concentration

Deepfakes tend to distract students from focusing because of their engaging, immersive, and sometimes sensational content, which can capture the attention of students and distract them from the essence of learning. Deepfakes capture the focus of students by making a hyper-realistic scenario or sensationalized version of lectures, fake news stories, or manipulated historical events.

With deepfakes, which can provide such very engaging visuals and audibles, there is the dependency for those students to be drawn away from study. It becomes less possible for them to focus on deep work because there exists a tendency to have the urge to watch entertaining or novel scenes.

On the other hand, deepfake technology, if applied in the right context of learning, can be a tool for innovative means of sustaining student engagement and concentration. For example, deepfakes can be used in creating personalized and interactive experiences that appeal to different kinds of learning styles.

3.2 Simulations

Through deepfake technology, teachers can use it to simulate historical figures, bring textbook contents alive, or present complex concepts in novel ways. This might engage students more because the content material would be more appealing and interactive, especially to those who are visual and auditory learners.

3.3 Critical Thinking

Deep fake would really pose great challenges in regards to critical thinking. Since deepfakes inherently alter the perception of truth, this brings fact and fiction under one cover. Hence, students become prone to losing critical thinking ability as far as whether any given information is actually true or just something put there.

3.4 Trust Issues

Deepfakes could break the trust of the students in any source be it academic or nonacademic. When a student becomes accustomed to convincing fake videos or audio recordings, he might develop a habit of questioning the validity of every source of information. This can be too broad, and students find it hard to identify credible sources or fail to critically evaluate authentic content.

3.5 *Susceptibility to Propaganda*

When applied with malicious intent, Deepfakes can mislead students in cases of political propaganda, disinformation campaigns, or fake academic credentials. Such exposure could lead to a decrease in the critical engagement with information and the questioning of biased or manipulated narratives. Students would be unable to engage in thoughtful, critical discussions or to form reasoned arguments without their ability to distinguish fact from fiction.

3.6 *Training to Detect Propaganda*

In their curricula, educators could include deepfake technology for their students to learn about manipulations of media and their subsequent need for critical thinking skills. Students can learn more about how media is put together and manipulated by scrutinizing deepfakes in great detail, which would thereby enrich their ability to critically understand content in the future. This means students could be assessed with evaluating a deep fake video or audio recording about the production of it, discussing whether and in which way it might shape the opinions of public in and upon society.

3.7 *Learning Lost in Entertainment*

If the students are constantly exposed to highly engaging but superficial content, such as deepfake videos, they would tend to love entertainment more than deep analytical thought. This might create shallow learning habits wherein the students take up easy-to-digest content rather than working with complex challenging material.

The ability to be able to engage deeply with complex texts or arguments may be reduced if students become accustomed to quick, visually stimulating content provided by deepfakes. Cognitive development relies on sustained engagement with challenging material, and if deepfakes are used to bypass the need for critical engagement, students might not develop the mental muscles needed to analyze and evaluate complex issues.

4 Benefits of Deep Fake in Higher Education

4.1 Versatile Learning Resource

History figures, scientists, and experts can be represented by making realistic simulations using deep fake technology. This will ensure that students are interacting in ways that make learning enjoyable and engaging. For example, a student could interview a digital avatar of Albert Einstein or receive a lecture from an expert philosopher reenacted for them.

4.2 Personalisation

The higher education is missing personalization. With deepfake technology, educators may personalize their video content based on learner requirements, such as breaking complex topics or delivering information differently depending on learner style preferences.

4.3 Disaster Management Training

Deepfakes can be used to simulate realistic disaster scenarios, and this may assist students in areas such as emergency management, medicine, or law enforcement in exercising critical decision-making in stressful situations.

4.4 Lowering Cost of Developing Content

Deepfake technology will also be able to repurpose and repackage educational videos, lectures, and other material to prevent wastage and decrease costs for new content while also keeping high-quality material available for longer periods.

4.5 Accessibility

Deep fake technology can be used to leverage available resources to reach out to a large number of students for better enrolment status of the higher education. Availability of smart phones in the world makes the accessibility feasible.

4.6 Academic Misconduct

It encompasses concerns of plagiarism and fraudulent work such that a student can apply the technology to impersonate professors or experts, even peers for them to dodge assignments or producing fake work. For example, they may pretend to discuss with an acclaimed academician or expert just to falsely show what they know about an issue.

4.7 Legal Issues

It would lead to the infringement of copyright and intellectual properties by creating deepfakes about educational content without the image of public figures, artists, or scholars. In any misuse of such technology in the courts of law, the institutes will face liability.

4.8 Lack of Human Touch

Deepfakes can simulate real people, but there is no way to get real human interaction, genuine empathy, or spontaneous communication with them. Over-reliance on virtual avatars and digital simulations may lead to a more boring, less personal education experience, which may in turn devalue the teacher student relationship.

4.9 Conclusion

The deepfake technology is developing rapidly, which creates a challenging problem for the education sector and calls for proactive regulatory action in dealing with issues of privacy, security, intellectual property, and ethics. While deepfakes have the potential to revolutionize educational experiences through improved engagement and accessibility, misuse can lead to massive harm, including misinformation, academic dishonesty, and violations of privacy. Educational institutions need to collaborate with lawmakers and regulatory bodies to create distinct, all-encompassing policies that would help protect deepfake technology from responsible, transparent misuse. By doing so, it may save the education system from dilution while limiting the emerging technology risks. Deepfakes' impact on the attention span of students and how critical thinking works for a student is a complex entwined combination of potential and challenge. Deepfakes may prove to be a very strong means for engagement and creative learning, hence enabling the students to better concentrate while providing novel ways to interact with the course material. However, deepfakes pose many risks to critical thinking. These create an erosion of a student's ability to analyze authenticity and can weaken media literacy; they also lead to shallow learning. Legislation, international cooperation, public awareness, and technological innovations in finding deepfakes and combat are a few of the features that are currently in the response. But growing advancement of deepfakes would mean that governments need to be agile and proactive in response to issues like manipulation, financial fraud, social harm, spreading hatred and anarchy.

References

1. Bhalli, N.N., Naqvi, N., Evered, C., Mallinson, C., Janeja, V.P.: Listening for Expert Identified Linguistic Features: Assessment of Audio Deepfake Discernment Among Undergraduate Students (2024). arXiv preprint arXiv:2411.14586
2. Black, J., Fullerton, C.: Digital deceit: fake news, artificial intelligence, and censorship in educational research. Open J. Soc. Sci. **8**(07), 71 (2020)
3. Bouderbane, A.: Using AI tools to prompt knowledge appropriately and ethically in higher education. Gen. Aspects Appl. Gener. AI High. Educ. Oppor. Chall. 327–345 (2024)
4. Caporusso, N.: Deepfakes for the good: a beneficial application of contentious artificial intelligence technology. In: Advances in Artificial Intelligence, Software and Systems Engineering: Proceedings of the AHFE 2020 Virtual Conferences on Software and Systems Engineering, and Artificial Intelligence and Social Computing, July 16–20, 2020, USA, pp. 235–241 (2021). Springer International Publishing.
5. Chesney, B., Citron, D.: Deep fakes: a looming challenge for privacy, democracy, and national security. Calif. L. Rev. **107**, 1753 (2019)
6. Deck, A.: Check for updates surveying impacts of AI in education and creative practices. In: Intelligent Systems and Applications: Proceedings of the 2023 Intelligent Systems Conference (IntelliSys) Volume 1, vol. 822, p. 400. Springer Nature (2024)
7. Delchev, K., Safieddine, F., Hammad, R.: Identification of AI generated deep fake video by higher education students. In: Science and Information Conference, pp. 473–489. Springer Nature Switzerland, Cham (2024)
8. Dumitrașcu, L.M., Lespezeanu, D.A., Zugravu, C.A., Constantin, C.: Perceptions of the impact of artificial intelligence among internal medicine physicians as a step in social responsibility implementation: a cross-sectional study. In: Healthcare, vol. 12, no. 15, p. 1502 (2024)
9. Ezzaim, A., Kharroubi, F., Dahbi, A., Aqqal, A., Haidine, A.: Artificial intelligence in education-State of the art. Int. J. Comput. Eng. Data Sci. (IJCEDS) **2**(2) (2022)
10. Greengard, S.: Will deepfakes do deep damage? Commun. ACM **63**(1), 17–19 (2019)
11. Ho, W., Woo, H.S., Lee, D.H., Kim, Y.: Development and distribution of deep fake e-learning contents videos using open-source tools. J. Distrib. Sci. **20**(11), 121–129 (2022)
12. Huber, C.R., Kuncel, N.R.: Does college teach critical thinking? A meta-analysis. Rev. Educ. Res. **86**(2), 431–468 (2016)
13. Lehto, M., Neittaanmäki, P.: Cyber security training in Finnish basic and general upper secondary education. In: The Proceedings of the... International Conference on Cyber Warfare and Security, vol. 18, no. 1. Academic Conferences International Ltd. (2023)
14. Liu, Y.: Research on the double-edged sword effect of AI in middle school students' learning applications. In: SHS Web of Conferences, vol. 209, p. 01015. EDP Sciences (2024)
15. Navarro Martínez, O., Fernández-García, D., Cuartero Monteagudo, N., Forero-Rincón, O.: Possible health benefits and risks of deepfake videos: a qualitative study in nursing students. Nurs. Rep. **14**(4), 2746–2757 (2024)

16. Rao, T.V.N., Koppula, M., Vardhan, G.H., Bhavana, S., Kumar, B.S.: The future of deepfakes: emerging trends and potential applications. Deepfakes Their Impact Bus. 243–266 (2025)
17. Seow, J.W., Lim, M.K., Phan, R.C., Liu, J.K.: A comprehensive overview of deepfake: generation, detection, datasets, and opportunities. Neurocomputing **513**, 351–371 (2022)
18. Sharma, R., Bhambri, P.: Digital duplicity and the disintegration of trust: a quantitative inquiry into the impact of deep fakes on media sustainability and societal equilibrium. In: Handbook of Technological Sustainability, pp. 273–291. CRC Press (2024)
19. Tiernan, P., Costello, E., Donlon, E., Parysz, M., Scriney, M.: Information and media literacy in the age of AI: options for the future. Educ. Sci. **13**(9), 906 (2023). https://doi.org/10.3390/educsci13090906
20. Ukata, P.F., Agburuga, V.: Application of Artificial Intelligence (AI) in teaching business education courses in rivers state tertiary institutions. Natl. Assoc. Bus. Educ. Niger. (2024)
21. Wiederhold, B.K.: Can deepfakes improve therapy? Cyberpsychol. Behav. Soc. Netw. **24**(3), 147–148 (2021)
22. Zhang, C., Lu, Y.: Study on artificial intelligence: the state of the art and future prospects. J. Ind. Inf. Integr. **23**, 100224 (2021)
23. Zhou, K.Z., Kilhoffer, Z., Sanfilippo, M.R., Underwood, T., Gumusel, E., Wei, M., Xiong, J.: Ethics, governance, and user mental models for large language models in computing education. XRDS: Crossroads, ACM Mag. Stud. **31**(1), 46–51 (2024)

The Impact of Intellectual Capital and Emotional Intelligence on Firm Performance of SMEs: The Mediating Role of Deep Fake Technology

Elif Hasret Kumcu, Rajat Sharma, and Muskan Singh

Abstract In today's competitive business environment, small and medium-sized enterprises increasingly require leveraging intangible assets like intellectual capital (IC) and emotional intelligence (EI) to gain a competitive edge. These are the critical drivers of success for an organization. Furthermore, the penetration of Deep Fake technology-an advanced artificial intelligence intended to create hyper-realistic digital representations in the organization-offers an extra dimension through which business operations can greatly enhance interactions with various stakeholders and customers. This chapter discusses how Emotional Intelligence (EI) and Intellectual Capital (IC) impact the Firm Performance of SMEs while showing the mediation effect of Deep Fake technology. Using the Partial Least Squares Structural Equation Modelling (PLS-SEM), the relationships among the various constructs as well as insights into how they generate success in SMEs will be discussed.

Keywords Deep-fake · Emotional Intelligence [EI] · Intellectual capital · Firm performance · SMEs · Technology · Artificial intelligence

1 Introduction

Historically regarded as fading even further, small and medium enterprises (SMEs) find themselves today barely able to thrive within uncertain conditions, the rapid development of technology, and trends on the international market [1]. To combat these difficulties, SMEs must arrange an ability to deal with the fast-evolving dynamics of the business environment by focusing not just on tangible assets but also intangible assets that build their competitive edge and success in the long term.

E. H. Kumcu
Aksaray University, Aksaray, Turkey
e-mail: elifhasretkumcu@aksaray.edu.tr

R. Sharma · M. Singh (✉)
Department of Management Studies, Graphic Era Deemed to Be University, Dehradun, India
e-mail: muskansinghofficial1998@gmail.com

S. Taneja et al. (eds.), *Navigating the Deepfake Conundrum: A Manager's Roadmap*, Information Systems Engineering and Management 44,
https://doi.org/10.1007/978-3-031-90742-5_14

There are two most recognized categories of intangible assets: intellectual capital and emotional intelligence, which are being considered these days to be very much important for good organizational performance [2]. Intellectual capital means the knowledge, experience, employees, and the pertinent intellectual property of an organization, while emotional intelligence means the ability to recognize, understand, and manage one's emotions and those of others. These factors have significantly contributed to the performance of a firm, especially in arenas where flexibility, innovation, and human interaction are sought from firm and its stakeholders.

The term intellectual capital is taken to mean, in a broad sense, the "knowledge-based assets" of a firm, which are the very essence required by SMEs for innovating, distinguishing themselves, and remaining competitive in an increasingly information-driven arena. Unlike conventional financial and physical assets, intellectual capital reflects the firm's ability to make use of knowledge, foster creativity, and then derive value from those very resources based on human skills and technology [3]. These include human capital (the skill, knowledge, and expertise of the employees), structural capital (the organizational processes and systems), and relational capital (the relationships with customers, suppliers, and other stakeholders). These components together serve as the backbone of the intellectual resources of a firm. The proper management and utilization of intellectual capital would enable SMEs to innovate, enhance decision-making processes, and gain a position of competitive advantage.

Sure, Emotional intelligence has not been studied like other influences, but it is also an important factor that affects SMEs performance. Examples of the capabilities that develop an individual's emotional intelligence are understanding oneself and others and motivational skills; empathy, self-control, self-regulation, social situation detection, and social management are all found in emotional intelligence. EI being less personal and direct, manifests itself in organizational culture, communication, and interpersonal relationships, both internally among the employees and externally with customers, suppliers, and stakeholders. Emotionally intelligent leaders and employees can better deal with conflicts, work in a team more effectively, and establish sustained customer relations-the sinews of success for SMEs [4].

While both IC and EI are important factors in assessing a small and medium-sized enterprise's performance, it cannot be understood in total independence from other factors. Rapid changes brokered by technology are also changing the nature of business stakeholder interactions and any other daily operation. One such technological novelty includes Deep Fake technology, which presents a more progressive AI capability that creates highly realistic but completely unreal people, objects, or events. Deep Fake is affirmed with issues of misinformation and privacy, but there is a unique opportunity for small-scale enterprises to develop customer engagement strategies, personalize marketing approaches, and make relationships more real with their audiences. This additional aspect of interaction over the digital medium can transform enormously the role of both intellectual capital and emotional intelligence in a business setting.

Deep Fake technology stands to be a really encouraging mediator interlinking the diverse ways through which SMEs employ intellectual capital and emotional intelligence for the purpose of stakeholder engagement and firm performance. For instance, one can imagine customized interaction in real-time for their customers through the use of AI-generated avatars or spokespersons, thereby improving customer satisfaction and loyalty through tailored communication with emotional intelligence [5]. The other area is; therefore, the organization can leverage Deep Fake technology to recreate and simulate creative scenarios and solutions, thus engaging their human capital's creative capabilities into a virtual environment. But still, the real problem is how these said emergent tools interact with existing intangible assets such as IC and EI and what the role of such tools is in enhancing firm performance.

In this chapter, we investigate the intricate nexus between intellectual capital, emotional intelligence, and Deep Fake technology. We also analyse the mediating effects of Deep Fake technology on the generation of firm performance by applying Partial Least Squares Structural Equation Modelling (PLS-SEM). This technique enables a study into the validity of the various interrelationships capturing their multidimensionality and worth in giving insights to the SMEs in their journey into digital transformation and the optimal use of their intangible assets [6]. The findings to be discussed will assist SMEs in making giant strides toward further enhancing organizational success and competitive advantage by combining fields of intellectual capital and emotional intelligence with emerging technologies such as Deep Fake in the current business climate.

The subsequent sections will analyse the theoretical meaning of intellectual capital and emotional intelligence, discuss the feasible applications of Deep Fake technology, and weigh the possible avenues of influence for these elements together on firm performance. The overarching goal of this analysis and discussion is to suggest practical recommendations and strategies for SMEs wishing to prosper in a technology-driven and ever-complex market.

2 Literature Review

In todays fast-paced and highly competitive business environment, small and medium enterprises are changing and keeping their pace to survive and thrive as organizations. They are increasingly under pressure, tangibly because of the growing information flow around businesses and intangibly because organizational success may increasingly depend on intellectual capital (IC) and emotional intelligence (EI). Intangible assets, therefore, empower SMEs in innovative ways to improve decision-making and build relationships with stakeholders [7]. This literature review discusses the roles of intellectual capital and emotional intelligence in the performance of SMEs and explores the mediation effects of integrating an advanced artificial intelligence form, Deep Fake technology, on these parameters and firm performance.

2.1 Intellectual Capital and Firm Performance

Intellectual capital (IC) has long been identified as a basic resource for organizations creating competitive advantages. Research describes intellectual capital as the intangible value held by organizations in knowledge, skills, experience, and proprietary information. It consists of three main components: human capital, structural capital, and relational capital [8]. Human capital refers to the knowledge, skills, and expertise of employees, while structural capital consists of the systems, processes, and culture facilitating the creation and sharing of knowledge. Finally, relational capital refers to the relationships and networks maintained by an organization with external stakeholders such as customers, suppliers, and partners.

Numerous studies have emphasized the significance of intellectual capital in SME performance. The research done, demonstrated that the effective management of intellectual capital would lead to an increase in innovation, productivity, and profitability. Small and medium enterprises that invest in knowledge-based assets will certainly have an advantage in adapting to the ever-changing dynamics of market demand since they utilize their human capital and structural systems for innovation and differentiation against competition rather than a well-articulated and defined capital. Since SMEs are generally more resource-constrained, they tend to rely more heavily on human capital. Thus, developing and retaining a talented workforce is crucial for achieving long-term success.

Additionally, relational capital is significant for SMEs that normally rely on strong ties with customers and suppliers for market access and trust development [9]. In most sectors, personalized service, reputation, and long-term partnerships are key success factors. Since SMEs usually lack financial and technological resources, be it non-MNCs' competitors, the ability to develop and maintain superior relationships presents a competitive advantage.

2.2 Emotional Intelligence and Firm Performance

One more intangible asset very much affecting firm performance is emotional intelligence (EI) especially in relation to SMEs. The research defined emotional intelligence as the ability to recognize, understand, and manage one's own emotions and, more importantly, those of others. Emotional intelligence covers the five important facets of self-awareness, self-regulation, motivation, empathy, and social skills. These emotional competencies could go a long way toward creating the conditions for effective communication, leadership, and teamwork, all of which have important ramifications for the success of an organization [10].

Many studies have examined the role of emotional intelligence in small and micro enterprises, emphasizing its impact on leadership, team dynamics, and client relationships. High-EI leaders tend to have a better understanding of the intricacies of interpersonal relationships and the resolution of conflicts with team motivation. This is especially true for SMEs, where leadership styles are more personal and face-to-face. Organizations where the leadership style is high in EI have increased employee engagement, job satisfaction, and organizational commitment, which ultimately makes a positive impact on business performance.

Equally significant is emotional intelligence in customer relations. In small to medium scale enterprises, where customer loyalty and trust are often directly forged, employees with higher EI empathically relate to the customer and give personalized services. This helps in deepening the relationship with customers and increasing customer satisfaction, which is a primary driver for long-term success. Findings reveal a direct correlation between emotional intelligence and customer satisfaction and retention, which subsequently improves SMEs' business performance.

2.3 *The Mediating Role of Deep Fake Technology*

Deep Fake technology, an advanced form of AI that allows for hyper-realistic digital representations, adds another level of value on top of SMEs' intellectual capital and emotional intelligence. It is a strong tool for personalizing communication, creating virtual representations of human emotions, and improving stakeholder engagement [11].

Realities cough up numerous applications for Deep Fake technology within the small-and-medium size enterprises realm. For instance, one application of this technology in marketing for SMEs would be personalizing advertisements through Deep Fake avatar or spokesperson who then engages a virtual customer in an emotionally intelligent manner. Through such advertisement delivery, an AI avatar could answer client queries live, adapt its tone to the current moods of the customer, and offer product suggestions customized to that customer's state. Such applications can economically help SMEs when combined with emotional intelligence to create experiences that increasingly become more real, emotionally engaging, thus creating closer bonds between customers.

Employee cooperation in collaborative environments for innovative brainstorming and knowledge-sharing creates intellectual property potential for Deep Fake technology. The technology can, therefore, simulate situations in which creativity and problem-solving are encouraged, forming a mechanism with which SMEs can fully utilize the potential of their human capital regardless of geographical and logistical constraining factors. In this regard, the virtualization of real-world scenarios can further enhance SMEs' decision-making and innovation processes.

With the widespread adoption of Deep Fake technology, it's imperative to consider its potential as a mediator between intellectual capital, emotional intelligence, and firm performance. For SMEs, the technology could facilitate the improvement of stakeholder engagement by utilizing intellectual and emotional resources to enhance communication, customer service, and general business performance [12]. However, more studies should be done to fully understand the mediating effect of Deep Fake technology with respect to SMEs, particularly in view of the interaction with other intangible assets like intellectual capital and emotional intelligence.

3 Research Methodology

This research is conducted to analyse the effects of Intellectual Capital (IC) and Emotional Intelligence (EI) on organizational performance of Small and Medium Enterprises (SMEs) in India; stressing on the mediators that Implicit Deep Fake technology. These all sorts of relationships will be examined by the Partial Least Squares Structural Equation Modelling (PLS-SEM), which is the major statistical technique for the purpose of data analysis. The forthcoming sections provide the research methodology used in the study for achieving the aim understudy- Research design, Sample selection, Data collection process and Analytical approach.

4 Research Design

This study has a research design and methodology that is both descriptive and quantitative. It is cross-sectional in that data is collected only once at a particular point in time for assessing the current state of intellectual capital, emotional intelligence, and firm performance in the Indian SME context. The study aims to establish relationships among these constructs and to examine mediation through Deep Fake technology using PLS-SEM for the simultaneous testing of many relationships within a complex model.

The research framework includes three primary constructs:

1. *Intellectual Capital (IC)*—This construct is evaluated through three dimensions: human capital, structural capital, and relational capital. Human capital represents the knowledge, skill, and abilities of employees themselves, while structural capital concerns the processes, systems, and culture of the organization. Relational capital, on the other hand, relates to the relationships that SMEs maintain with such external stakeholders as customers, suppliers, and partners.
2. *Emotional Intelligence (EI)*—The operational definition of emotional intelligence was built on the five dimensions mainly conceived through research: namely self-awareness, self-regulation, motivation, empathy, and social skills.
3. *Firm Performance*—The firm performance is evaluated using a set of subjective and objective indicators such as financial success, increased market share, customer satisfaction, and innovation results.
4. *Deep Fake Technology*—Examination of Deep Fake technology as a mediating variable suggests that its operation might influence the relationship between IC, EI, and firm performance. The assessment of the role of Deep Fake technology

as a means of enhancing stakeholder engagement and communication within the context of marketing, customer service, and innovation is hereby undertaken.

4.1 Sample Selection

The sample for this study consists of 70 SMEs based in India, selected based on the criteria that they operate for at least two years, represent a wide range in terms of industries, e.g., manufacturing, services, retail, and technology, and include employees ranging between 50 and 250. These SMEs are considered representative of the Indian SME sector, which varies in size, industry, and business strategies.

Moreover, sample is from non-probability convenience sampling method. This is because available data are relatively few, and practicalities attached to getting a larger and more heterogeneous sample are cumbersome. However, convenience sampling does not ensure a perfectly representative sample, but it affords the research access to adequate data from respondent firms which are most likely to provide information useful in understanding the effect of intangible assets on performance level.

4.2 Data Collection

It would be best to get information from senior officials or owners of the SMEs selected for this study through a structured questionnaire. This gives insight into the information necessary with respect to constructs- intellectual capital, emotional intelligence, firm performance, and usage of Deep Fake technology. These were adapted from already established scales in literature to ensure that validity in terms of content and reliability is satisfied.

The questionnaire consists of the following sections:

- Section 1: Demographic Information—A collection of basic information, with industry type, size, and years in operation, is detailed.
- Section 2: Intellectual Capital—Items regarding three forms of intellectual capital: human, structural, and relational capital (adapted from [13, 14]).
- Section 3: Emotional Intelligence—Items measuring the five components of EI (self-awareness, self-regulation, motivation, empathy, and social skills) [15, 16].
- Section 4: Firm Performance—The indicators of company performance consist of a mixture of objective and subjective performance indicators viewed through a prism of financial performance, growth, and innovation [17].
- Section 5: Deep Fake Technology—Queries regarding the extent to which businesses use Deep Fake Technologies for their marketing, customer interactions, and innovation processes (adapted from [18]).

4.3 Analytical Approach

To analyze the relationships between IC, EI, Deep Fake technology, and firm performance, Partial Least Squares Structural Equation Modelling (PLS-SEM) is employed. PLS-SEM is chosen due to its ability to handle complex models, assess causal relationships, and analyze latent variables with relatively small sample sizes. The analysis consists of two stages:

- **Measurement Model Assessment**: This stage evaluates the reliability, validity, and discriminant validity of the constructs to ensure that the indicators accurately measure the underlying variables.
- **Structural Model Assessment**: This stage tests the hypothesized relationships among IC, EI, Deep Fake technology, and firm performance by examining path coefficients, effect sizes, and mediation effects.

5 Data Analysis and Discussion

From the above analysis it can be concluded that human capital, relational capital and emotional intelligence significantly boost firm performance while structural capital which is the third factor of intellectual capital does not effectively contribute to firm performance growth. Moreover from the analysis it can be concluded that deepfake technology significantly act as a mediator and affects firm performance (Fig. 1 and Table 1).

Deep Fake (0.798): This indicates good internal consistency, meaning that the items related to "Deep Fake" are reliably measuring the same construct.

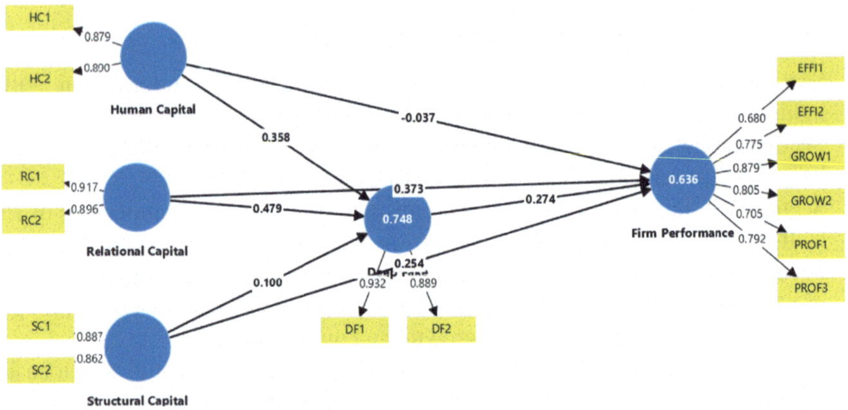

Fig. 1 PLS SEM analysis graphical representation

Table 1 Construct reliability and validity table

Variables	Cronbach's alpha	Composite reliability (rho_a)	Composite reliability (rho_c)	Average variance extracted (AVE)
Deep Fake	0.798	0.828	0.907	0.830
Firm Performance	0.866	0.875	0.900	0.602
Human Capital	0.722	0.723	0.878	0.783
Relational Capital	0.784	0.790	0.902	0.822
Structural Capital	0.693	0.697	0.867	0.765

- Firm Performance (0.866): This is also a strong value, suggesting a very reliable measure for firm performance.
- Human Capital (0.722): This is acceptable, though slightly lower than the optimal range.
- Relational Capital (0.784): This value shows good reliability, similar to Deep Fake.
- Structural Capital (0.693): This is on the lower end, slightly below the acceptable threshold (0.7), which might suggest that the scale could benefit from some improvements in reliability.

EXPLANATION FOR AVE

- Deep Fake (0.830): This is a good score, indicating that the "Deep Fake" construct explains a large portion of the variance in its indicators.
- Firm Performance (0.602): This is a bit lower than ideal. Ideally, you would want this to be higher (above 0.7), suggesting that the measurement model for firm performance may not be fully capturing the underlying construct.
- Human Capital (0.783): This is a good score, meaning the "Human Capital" construct explains a good portion of the variance in its items.
- Relational Capital (0.822): This is a strong score, suggesting good convergence of the items with the construct.
- Structural Capital (0.765): This is a good score as well, indicating that the structural capital measure is capturing most of the variance in the items (Table 2).

Table 2 R square table

Variable	R-square	R-square adjusted
Deep Fake	0.748	0.736
Firm Performance	0.636	0.613

- Deep Fake (0.748): This R^2 value means that the independent variables (predictors) in the model explain about 74.8% of the variance in the Deep Fake variable. This is a strong explanatory power, suggesting that the predictors are quite effective in explaining the outcomes related to Deep Fake.
- Firm Performance (0.636): This means that the model explains about 63.6% of the variance in Firm Performance. This is still a relatively strong result, indicating that a significant portion of the variation in firm performance is explained by the independent variables in the model.

6 Conclusion and Limitations of the Study

The modern-day business environment is fast changing, and for SMEs to improve their performance and sustain a competitive edge, they would have to rely heavily on the development of intangible assets: Intellectual Capital (IC) and Emotional Intelligence (EI). In this study, IC, and EI's considerable influences on Firm Performance are examined, analysis being made of how these factors stimulate innovation, enhance decision-making, and strengthen relationships with stakeholders.

In addition, the Deep Fake technologies act as a significant mediator and provide SMEs with new means to communicate with customers, personalize their marketing, and improve operations. The results from the Partial Least Squares Structural Equation Modelling (PLS-SEM) analysis indicated that Deep Fake technology can intensify the advantages of IC and EI, thus generating ultimately improved business outcomes.

With digital transformation steadily transitioning industries, SMEs will do well to invest strategically in such intangible assets while adopting emergent technologies powered by AI. Such investments increase their agility, enhance sustainable growth, and confer long-term success in an ever-growing competitive environment. Future works may discuss ethical considerations and the application of industry-specific measures to optimize Deep Fake technology for gains while limiting the unwanted perils associated with its use.

On the verge of being helpful, this study has its limitations that need consideration when interpreting its findings. One limiting factor could be the sample size of 70 SMEs, though perhaps enough for the analysis. A larger and more diverse sample, with SMEs from different geographic regions and industries, would increase robustness and enhance our understanding of the dynamics involved in this study. Indian SMEs may thus possess limitations in their cross-cultural relevance. The different business environments, problems, and technology adoption (in this case, Deep Fake) in different countries and regions will decide how far the findings can be applied to SMEs elsewhere in the world.

This research, second in line, is a cross-sectional one, data collection for which occurred at a single point in time. This very somewhat restricts the inference concerning any causal relationships concerning the constructs, as changes over time or external factors affecting these changes may not be valid accounting options. A longitudinal study would have been useful in further assessing how the relationships between the constructs change over time and on a broader aspect of how Deep Fake technology adoption transversely impacts firm performance.

While this study has focused on intellectual capital, emotional intelligence, and Deep Fake technology, other elements influencing SME performance—such as market conditions, access to finance, and regulatory challenges—were not explored. Future studies may consider other variables in the scope of the study from a more holistic perspective on factors influencing SME success. So, while this study makes a significant contribution to the understanding of the role of intangible assets and emerging technologies in the performance enhancement of SMEs, some further studies with a larger sample size and longitudinal design would lend more credence to, and extend, the findings already made.

References

1. Agarwal, R., Gupta, S.: Facades and fortunes: intellectual capital's influence on firm dynamics in the deepfake epoch. ResearchGate (2024). Retrieved from https://www.researchgate.net
2. Bennet, M., Zhao, H.: The impact of AI deepfakes on human cognition and emotion. Preprints (2023). Retrieved from https://www.preprints.org
3. Cheng, Y., Adams, P.: Prediction of self-efficacy in recognizing deepfakes based on personality traits. Psychiatr. Cogn. Neurosci. **12**(4), 87–102 (2023). Retrieved from https://pmc.ncbi.nlm.nih.gov
4. Davis, K., Lin, C.: Research on the effects of deepfake technology for the modern digital space. J. Digit. Ethics **14**(2), 56–73 (2022). Retrieved from https://www.researchgate.net
5. Evans, R., Patel, J.: After deception: how falling for a deepfake affects the way we see others. Journal. Mass Commun. Quart. (2024). Retrieved from https://journals.sagepub.com
6. Foster, L.: Implications of deepfake technology on individual privacy and security. St. Cloud State University Research Repository (2023). Retrieved from https://repository.stcloudstate.edu
7. Gonzalez, P., Kim, S.: The public mental representations of deepfake technology. Cogn. Sci. J. **27**(3), 112–128 (2023). Retrieved from https://pmc.ncbi.nlm.nih.gov
8. Henderson, B.: Analyzing the legal and ethical considerations of deepfake technology. Int. Stud. Law Policy **18**(1), 22–39 (2023). Retrieved from https://journalisslp.com
9. Johnson, R., Lee, A.: Audio deepfake: challenges and future directions. Wikipedia (2024). Retrieved from https://en.wikipedia.org
10. Nguyen, Q.V., Tate, M., Calvert, P., Aubert, B.: Leveraging ERP implementation to create intellectual capital: the role of organizational learning capability. J. Knowl. Manag. **20**(3), 653–677 (2016)
11. Ramos-Rodríguez, A.R., Medina-Garrido, J.A., Lorenzo-Gómez, J.D., Ruiz-Navarro, J.: What you know or who you know? the role of intellectual and social capital in opportunity recognition. J. Bus. Ventur. **39**(1), 106001 (2024)
12. Hasyyati, A.N., Kurniawan, A.: Measuring intangible assets using parametric and machine learning approaches. J. Econ. Dyn. Control **144**, 104224 (2022)
13. Bontis, N.: Intellectual capital: an exploratory study that develops measures and models. . Manag. Decis. **36**(2), 63–76 (1998). https://doi.org/10.1108/00251749810204142
14. Stewart, T.A.: Intellectual Capital: The New Wealth of Organizations. Doubleday/Currency (1997)
15. Goleman, D.: Emotional Intelligence: Why It Can Matter More Than IQ. Bantam Books (1995)

16. Schutte, N.S., Malouff, J.M., Thorsteinsson, E.B., Bhullar, N., Rooke, S.E.: A meta-analytic investigation of the relationship between emotional intelligence and health. Pers. Individ. Differ. **47**(2), 921–933 (2009). https://doi.org/10.1016/j.paid.2009.07.014
17. Venkatraman, N., Ramanujam, V.: Measurement of business performance in strategy research: a comparison of approaches. Acad. Manag. Rev. **11**(4), 801–814 (1986). https://doi.org/10.5465/amr.1986.4283976
18. Chesney, R., Citron, D.K.: Deepfakes and the new disinformation war: the coming age of post-truth geopolitics. Foreign Aff. **98**(1), 147–155 (2019)

Deconstructing Deception: The Legal, Ethical, and Technological Nexus of Deepfakes in G7 Business Practices

Swati Gupta and Ercan Ozen

Abstract The rapid advancement of deepfake technology, driven by sophisticated artificial intelligence algorithms such as Generative Adversarial Networks (GANs), has introduced unprecedented challenges to business practices, particularly within the G7 nations. Deepfakes—hyper-realistic digital forgeries of audio, video, and imagery—have the potential to undermine corporate integrity, erode public trust, and destabilize financial markets. This study explores the legal and ethical implications of deepfakes, with a focus on their disruptive impact on business operations in these technologically advanced countries. Through a comprehensive literature review, secondary data analysis, and case studies, the research identifies the key risks associated with deepfakes, including fraud, defamation, and misinformation. It critically examines existing regulatory frameworks, highlighting the insufficiency of current laws to address the complex realities of digital impersonation and identity theft. Furthermore, the ethical concerns surrounding the manipulation of corporate communications and consumer perception are explored, emphasizing the necessity for enhanced transparency and accountability in business practices. The study concludes by advocating for the development of robust international legal standards, advanced cybersecurity protocols, and ethical guidelines to mitigate the risks posed by deepfake technology. As deepfakes continue to evolve, businesses must adapt to an increasingly uncertain digital environment, balancing innovation with responsibility to maintain credibility, trust, and security.

Keywords Deepfake technology · Artificial intelligence · Generative adversarial networks · Business ethics · Corporate governance · Legal frameworks · Digital security · Misinformation

S. Gupta (✉)
Chitkara Business School, Chitkara University, Punjab, India
e-mail: swati.gupta@chitkara.edu.in

E. Ozen
Usak University, Usak, Turkey
e-mail: ercan.ozen@usak.edu.tr

S. Taneja et al. (eds.), *Navigating the Deepfake Conundrum: A Manager's Roadmap*, Information Systems Engineering and Management 44,
https://doi.org/10.1007/978-3-031-90742-5_15

1 Introduction

In the age of technological ubiquity and digital manipulation, the advent of deepfakes stands as a monumental paradigm shift, implicating multifaceted ethical, legal, and societal dimensions. These AI-generated hyper-realistic forgeries of audio, video, and images have proliferated with unprecedented velocity, presenting a new frontier in the complex nexus of digital innovation and regulation. The convergence of artificial intelligence (AI), machine learning (ML), and data-driven technologies has given rise to deepfakes, which enable the creation of convincingly altered multimedia content. Such content can mimic voices, replicate faces, and distort reality with such fidelity that discerning authenticity from fabrication has become an increasingly formidable challenge [49]. As these technologies mature, they present an evolving threat to trust, integrity, and accountability within business and corporate practices, especially within the Group of Seven (G7) nations [106].

The G7, comprising Canada, France, Germany, Italy, Japan, the United Kingdom, and the United States, represents some of the most advanced economies in the world. These nations, with their robust digital infrastructures, sophisticated market mechanisms, and powerful global influence, are simultaneously the most susceptible to the disruptive potential of deepfakes [128]. For businesses operating in these countries, the ramifications are far-reaching, ranging from brand damage to potential violations of corporate governance and compliance standards. The ability of deepfakes to fabricate misleading corporate communications, impersonate executives, manipulate financial data, and orchestrate disinformation campaigns presents unique challenges that transcend traditional cybersecurity concerns. Thus, it is imperative to consider the legal and ethical frameworks required to address these emerging risks [62].

Deepfake technology hinges upon complex algorithms, particularly generative adversarial networks (GANs), which have revolutionized the capabilities of artificial intelligence. GANs create hyper-realistic outputs by pitting two neural networks against each other, effectively mimicking the nuanced elements of human behavior, voice modulation, and visual identity. This technological prowess enables the creation of synthetic representations that are indistinguishable from authentic content, resulting in a host of potential exploitations that challenge the very notion of digital trust [69].

Within the realm of business practices, the consequences of deepfakes can be manifold. From an ethical standpoint, deepfakes introduce a host of moral dilemmas surrounding authenticity, transparency, and the integrity of corporate communications [146]. Executives, stakeholders, and customers rely on the credibility of visual and auditory content to make informed decisions about financial investments, product development, and corporate governance. The ease with which deepfakes can be deployed to manipulate these perceptions undermines the foundational trust upon which business ecosystems are built.

Moreover, deepfakes have the potential to exacerbate existing vulnerabilities in the corporate governance framework. Consider, for example, the use of deepfake technology to manipulate shareholder meetings, distort financial disclosures, or impersonate senior executives in order to execute fraudulent transactions or manipulate stock prices [110]. Such actions, if left unchecked, could lead to significant economic destabilization, loss of public confidence, and reputational damage to companies that fall victim to these sophisticated digital schemes.

Legally, the integration of deepfakes into business practices raises profound questions about intellectual property, data privacy, and corporate liability. In jurisdictions like the United States and the European Union, where legal protections for digital content and privacy are increasingly stringent, the use of deepfake technology to create fraudulent corporate materials could result in violations of intellectual property law, consumer protection statutes, and data privacy regulations [105]. The misrepresentation of individuals through synthetic media can also violate laws related to defamation, identity theft, and harassment, further complicating the legal landscape.

The ethical and legal ramifications of deepfakes also intersect with broader concerns about misinformation and disinformation in the digital age. In an era where corporate brands and public perception are often shaped by online narratives, the proliferation of deepfake content could foster a fertile ground for corporate espionage, malicious political agendas, and subversive tactics aimed at destabilizing markets or tarnishing reputations [145]. The resulting erosion of trust between businesses, consumers, and investors could have long-lasting consequences on the integrity of markets, thereby necessitating robust regulatory frameworks to safeguard against the misuse of deepfakes.

Internationally, the G7 countries are not only bound by domestic legal standards but also influenced by a complex web of transnational agreements and multilateral initiatives aimed at combating digital malfeasance. As such, the implications of deepfakes for business practices must be analyzed within a global context, taking into account divergent legal systems, regulatory priorities, and cultural attitudes towards digital content manipulation [16]. While some G7 nations have initiated legislative measures to address the risks posed by deepfakes—such as the implementation of deepfake detection technologies, the regulation of synthetic media, and the creation of penalties for malicious use—others remain ambivalent, opting instead for a more laissez-faire approach.

One of the key challenges for businesses in G7 countries is the regulatory uncertainty surrounding deepfakes. As a nascent technology, deepfakes are not yet adequately addressed by existing legal frameworks. For instance, while copyright law provides certain protections against unauthorized use of an individual's likeness or voice, deepfake technology complicates the application of these legal principles [11]. The rapid evolution of AI technology outpaces the ability of lawmakers to draft precise legislation, leaving businesses to navigate a murky landscape where legal protections are often inconsistent or ill-defined. Consequently, businesses are left vulnerable to emerging risks that may be difficult to predict, prevent, or litigate.

Furthermore, the ethical obligations of companies to protect their stakeholders from the potential harms of deepfakes are becoming increasingly pronounced. In an era of heightened social consciousness, businesses are held to higher standards of accountability not only by regulators but by the public as well [48]. As such, there is growing pressure for corporations to take proactive measures in addressing the risks associated with deepfakes, whether through internal cybersecurity protocols, employee training, or public awareness campaigns. Failure to act decisively could lead to irreparable harm to a company's brand equity and consumer trust, with potentially catastrophic financial implications.

Given these multifaceted challenges, the need for a cohesive and comprehensive legal and ethical framework to address the use of deepfakes in business practices within the G7 nations is paramount. This framework must not only address the immediate risks posed by deepfakes but also anticipate future developments in AI and digital content manipulation [8]. Policymakers must balance the imperatives of technological innovation with the need for robust safeguards that protect the public, businesses, and markets from the potential misuse of deepfake technologies.

In the subsequent sections of this paper, we will explore the specific legal and ethical implications of deepfakes for business practices in G7 countries. We will examine the intersection of deepfakes with intellectual property law, privacy rights, and corporate governance, while also considering the broader societal impacts of deepfakes, including their role in the erosion of trust and the amplification of misinformation. Through this analysis, we aim to provide a comprehensive understanding of the challenges posed by deepfakes and propose actionable recommendations for businesses and regulators to mitigate these risks in an increasingly digital and interconnected world.

2 Literature Review

The emergence of deepfakes has not only shaken the foundations of digital media but has also stirred significant discourse regarding their legal and ethical ramifications, particularly within the business sector. These sophisticated forgeries, enabled by advanced artificial intelligence (AI) techniques, present an imminent threat to the authenticity and integrity of digital content. The potential for deepfake technology to manipulate public perception, deceive consumers, and undermine corporate governance is profound, yet the frameworks to address these challenges are still in nascent stages [6, 125]. As such, a comprehensive review of the literature surrounding deepfakes' implications for business practices within the G7 nations reveals a confluence of technological, ethical, legal, and socio-political considerations [91].

At the core of the discourse on deepfakes lies the technological innovation itself. Deepfakes are primarily the result of generative adversarial networks (GANs), a class of machine learning algorithms that have revolutionized the ability to manipulate digital media with an unprecedented degree of realism. GANs function by pitting two neural networks against one another: one generates synthetic media, while the other evaluates the authenticity of the content, continuously refining the output until it is indistinguishable from the original [67, 104]. This dual-process architecture enables the creation of hyper-realistic videos, audio clips, and images, often making it impossible for the average observer to discern falsified content from genuine material. The rapid advancement in GAN capabilities has blurred the line between fact and fiction, raising concerns about the increasing reliance on digital media for decision-making processes in both the public and private sectors.

The legal implications of deepfake technology are complex and multifaceted, particularly for businesses operating in the regulatory environments of G7 countries. Legal scholars and practitioners have emphasized the necessity of creating new legal frameworks to address the challenges posed by deepfakes. Existing laws—such as those governing defamation, fraud, and intellectual property—are often insufficient in dealing with the nuances of digital content manipulation. The concept of liability for deepfake creation and dissemination remains an ambiguous area of law [4, 39]. Who, for instance, should be held accountable when deepfakes are used to impersonate high-level executives, manipulate financial records, or distort public communications? In many cases, perpetrators of deepfake-related fraud may remain anonymous, utilizing decentralized platforms and encrypted channels to distribute harmful content. Furthermore, as deepfake technology often involves the unauthorized use of an individual's likeness or voice, issues of privacy and intellectual property infringement are intertwined with broader concerns about digital identity theft.

Despite the clear need for legislative intervention, there exists considerable resistance to the establishment of laws specifically targeting deepfakes. Critics argue that overly stringent regulations could stifle innovation, particularly in the burgeoning fields of AI and machine learning. They contend that the capacity for AI-generated content to enhance creative industries—such as entertainment, advertising, and marketing—should not be undermined by overly restrictive legal constraints [33, 51]. Moreover, there is the challenge of balancing freedom of expression with the protection of individuals and businesses from harm. The legal literature on deepfakes thus grapples with the tension between fostering innovation and protecting the rights of individuals and institutions from malicious or fraudulent use of AI technologies.

In parallel with the legal concerns, the ethical dimensions of deepfakes in business contexts present another significant layer of complexity. The ethical implications of deepfake technology are vast, as the ability to manipulate reality threatens the trust and transparency that are foundational to business relationships. In particular, businesses that rely on public trust, such as financial institutions, technology companies, and media conglomerates, are particularly vulnerable to the potential consequences of deepfakes. The use of deepfakes to impersonate executives, alter financial data, or spread misinformation can not only harm the reputation of a company but also erode the trust that is essential for long-term success in the marketplace.

Ethicists have raised concerns about the potential for deepfakes to exacerbate the already widespread phenomenon of digital disinformation. Within the context of business, disinformation can take on particularly nefarious forms, such as false endorsements, fabricated product reviews, or even staged corporate events [26, 129]. These manipulations can deceive stakeholders, mislead investors, and undermine corporate strategies, leading to significant financial losses and damage to brand equity. The ability of deepfakes to mimic real individuals and organizations at a granular level complicates efforts to discern truth from deception, thus raising questions about the ethical responsibility of companies to safeguard against such technological abuses.

At the heart of the ethical dilemma is the question of consent. In many instances, deepfakes are created using the likeness or voice of an individual without their permission, thus violating their personal autonomy and privacy rights [130, 141]. In a corporate context, this could manifest as the unauthorized use of an executive's image to promote a product or service, or the falsification of business communications using a CEO's voice to authorize fraudulent transactions. The ethical principle of informed consent is integral to addressing these concerns, as businesses must establish mechanisms for protecting the rights of individuals whose likenesses are used in digital media.

Furthermore, the ethical quandary surrounding deepfakes is compounded by the rapid pace of technological development. As AI and machine learning algorithms continue to evolve, the capacity for deepfakes to deceive and manipulate expands in tandem [71, 131]. This rapid evolution presents significant challenges for businesses and policymakers alike, as the regulatory landscape struggles to keep pace with the speed of technological advancements. As deepfake technology becomes more accessible and sophisticated, there is an increasing need for businesses to implement proactive measures to detect and counteract digital forgeries before they can cause irreparable harm.

The ethical implications are further exacerbated by the global nature of the digital economy. In the interconnected world of the G7 nations, businesses are subject to a diverse array of regulatory environments, each with its own set of rules governing digital content, privacy, and intellectual property. However, the globalized nature of online platforms—combined with the decentralized and pseudonymous nature of deepfake creators—poses significant challenges for cross-border regulation [31, 144]. The difficulty of enforcing national laws in a borderless digital landscape complicates efforts to combat the malicious use of deepfakes in business contexts. This is particularly true when perpetrators are located in jurisdictions with weak or non-existent regulations surrounding digital content manipulation.

Additionally, the societal implications of deepfakes within business practices cannot be overlooked. As deepfakes continue to proliferate, they have the potential to contribute to the further erosion of public trust in digital media. In an age where misinformation and disinformation are already pervasive, deepfakes add a new layer of complexity to the problem of digital literacy. Consumers and stakeholders must become increasingly adept at discerning the veracity of the content they encounter, a skill that is particularly difficult given the increasing realism of AI-generated media [85, 139]. In the business sector, this poses a significant challenge, as stakeholders—including investors, consumers, and employees—must contend with the growing uncertainty surrounding the authenticity of corporate communications.

As the G7 nations continue to grapple with these complex legal and ethical questions, it is clear that business practices will need to evolve in order to adapt to the new realities of the digital age. Companies will need to adopt robust internal policies

[10, 40], invest in cutting-edge technologies to detect and mitigate the risks posed by deepfakes, and work with governments to ensure that regulatory frameworks are in place to address the rapidly changing landscape of digital content manipulation. In doing so, businesses must strike a delicate balance between embracing technological innovation and safeguarding the ethical principles that underpin their operations [98, 99].

The ongoing discourse surrounding the legal and ethical implications of deepfakes in business practices underscores the necessity for a multifaceted approach to address these emerging risks. Through the development of more comprehensive legal frameworks, the establishment of ethical guidelines, and the investment in technological solutions, businesses can better position themselves to navigate the complexities of this new digital frontier [5, 17, 138]. As deepfake technology continues to evolve, so too must our understanding of its implications, ensuring that the digital landscape remains a space of trust, transparency, and accountability.

In conclusion, the literature on the legal and ethical implications of deepfakes in business practices within the G7 countries reveals a profound and evolving challenge. As deepfake technology advances, its potential for misuse in business contexts grows, necessitating an urgent and nuanced approach to regulation and ethical conduct [80, 102, 127]. Businesses, policymakers, and stakeholders must work collaboratively to develop frameworks that mitigate the risks of deepfakes while fostering innovation and protecting the integrity of digital content. The stakes are high, and the need for comprehensive solutions has never been more pressing.

The relevant findings from past literature is shown in Table 1.

3 Research Methodology

The research methodology employed in this study is an amalgamation of qualitative analysis and secondary data synthesis, meticulously designed to explore the legal, ethical, and technological ramifications of deepfake technology within the business practices of G7 countries. The investigative approach is grounded in a thorough literature review, encompassing both academic and industry sources, which provides a comprehensive understanding of the technological underpinnings of deepfakes, their socio-economic implications, and the evolving landscape of regulatory frameworks. A combination of case studies and statistical data, drawn from governmental reports, cybersecurity agencies, and financial institutions, is employed to assess the prevalence, impact, and response strategies to deepfake-related incidents within these countries. This methodology is complemented by a comparative analysis of the legal doctrines and ethical frameworks across the G7 nations, enabling a nuanced evaluation of how these countries are adapting—or failing to adapt—to the disruptive capabilities of deepfake technology. The study also integrates a critical examination of emerging trends and predictive models, where available, to ascertain the future trajectory of deepfake-related challenges in business governance and digital security.

Table 1 Related studies

Study/author	Focus area	Key findings	Implications for business	Legal and ethical issues
Technological foundations of deepfakes [140]	AI and generative adversarial networks (GANs)	Deepfake technology is powered by GANs, which pit two neural networks against each other to create convincing media. The output is indistinguishable from real content	Businesses are at risk from manipulation of multimedia content such as advertisements, corporate communications, and interviews	Legal challenges include defining liability for AI-generated content. Ethical concerns revolve around misrepresentation
Liability and accountability in digital content manipulation [37]	Liability and accountability frameworks for digital content	Deepfakes blur lines of accountability. It is unclear whether the creator, platform, or end-user should be held responsible for harmful deepfakes	Corporate executives could be impersonated for fraudulent actions, such as signing contracts or making financial decisions	Ethical dilemmas surround whether businesses can ethically use AI for content creation and prevent misuse
Data protection and privacy implications [126]	Privacy laws and data protection in the age of deepfakes	Deepfakes can violate personal privacy, especially when using real individuals' likeness without consent	Corporate use of personal data, such as influencer marketing, could be compromised by unauthorized deepfakes	Violation of privacy rights and concerns about the misuse of individuals' likeness for unauthorized commercial purposes
Business reputation and brand damage [119]	Brand protection and corporate reputation	Deepfakes have the potential to damage a company's brand image, especially when fake videos or statements are attributed to key figures within the organization	The potential for brand sabotage is high—competitors or malicious actors could use deepfakes to harm a business's reputation	Ethical issues around reputation management, including whether businesses are responsible for guarding against digital forgeries

(continued)

Table 1 (continued)

Study/author	Focus area	Key findings	Implications for business	Legal and ethical issues
Financial market manipulation and fraud [19]	Financial fraud and market manipulation using deepfakes	Deepfakes could manipulate stock prices, with false news reports or CEO statements causing significant market fluctuations	Business executives and stakeholders must be vigilant in the face of digital manipulation affecting financial markets	Legal frameworks for fraud and market manipulation need to be adapted to include digital forgeries. Ethical concerns about market fairness arise
Intellectual property concerns [56]	IP protection and counterfeiting	Deepfake technology enables counterfeit production of content, infringing on intellectual property (IP) rights	Intellectual property rights, especially in marketing and advertising, are under threat from counterfeit deepfakes	Ethical concerns include the misuse of copyrighted materials, while legal protections for digital media are currently insufficient
Disinformation and corporate communications [150]	Corporate communications and the risk of disinformation	Deepfakes can create highly believable misinformation, potentially altering consumer perception or public opinion about businesses	Disinformation could lead to consumer mistrust, affecting purchasing decisions and brand loyalty	Businesses must decide on the ethical use of AI to prevent or expose disinformation. Legal responsibility for disinformation needs clarification
Ethical AI use in business [77]	Ethics in AI-driven business practices	While AI holds transformative potential, its misuse through deepfakes challenges ethical standards for honesty and transparency in business communications	Companies must balance innovation with ethical responsibility, ensuring AI usage is transparent and aligns with ethical norms	The ethical dilemma of creating AI that respects privacy and transparency, ensuring it isn't used to deceive or exploit

(continued)

Table 1 (continued)

Study/author	Focus area	Key findings	Implications for business	Legal and ethical issues
Governance and regulatory response [47]	Regulatory approaches to deepfake technology	The absence of a unified international legal framework leaves businesses vulnerable to inconsistent legal standards and regulatory compliance issues	Corporate leaders must stay abreast of emerging regulations, as inconsistent legal responses in different G7 countries could create compliance challenges	The global governance of AI technologies is inconsistent, requiring coherent policies to mitigate risks across borders
Public perception and trust in business [22]	Consumer trust in digital media and its impact on businesses	Deepfakes pose a significant threat to public trust in digital media, with businesses potentially facing reputational crises if consumers believe false content	Trust in businesses may erode if deepfake-related scandals become common. Businesses must work to rebuild consumer confidence	Ethical considerations around how businesses manage consumer trust in the face of emerging digital manipulations

4 Data Analysis

Below is a tabular representation in Table 2 of the history of technological advancements in G7 countries, highlighting key innovations, milestones, and their impact.

Table 3 contains recent developments in G7 countries, highlighting key events, their significance, and potential impacts.

Table 4 is summarising statistics related to deepfakes in G7 countries, focusing on their prevalence, impact, and related issues.

The proliferation of deepfake technology presents a multifaceted array of legal and ethical challenges, particularly within the context of business practices in the G7 countries. At its core, deepfakes exploit advanced generative algorithms, such as Generative Adversarial Networks (GANs), to create hyper-realistic digital forgeries of individuals' voices, faces, and behaviors. This capability introduces profound risks to business integrity, fraud prevention, and public trust, posing significant legal questions surrounding liability, intellectual property, privacy, and defamation [27, 70, 101]. One of the most pressing legal concerns is the question of accountability—who bears responsibility when a deepfake is used maliciously to undermine corporate reputation or manipulate financial markets? In many jurisdictions, including the G7, existing laws are ill-equipped to address the nuances of digital impersonation and

Table 2 Chronology of technological advancements in G7 countries: a pioneering legacy

Country	Technological advancement	Year	Key Milestones/ innovations	Impact on society and economy
United States	Invention of the internet [79, 149]	1960s–1990s	Developed by DARPA, the ARPANET evolved into the modern internet. Commercialization began in the 1990s	Revolutionized communication, business models, and global connectivity. Created entire industries (e-commerce, social media, cybersecurity)
United States	Silicon valley and personal computing [93, 113]	1970s–1980s	Rise of Silicon Valley as the hub for personal computers with innovations from companies like Apple, Microsoft, and Intel	Led to the digital age, democratized computing power, and fueled the global IT industry, including software, hardware, and innovation ecosystems
Canada	Development of 5G wireless technology [7, 28]	2010s	Canada contributed significantly to the research and testing of 5G networks, with key contributions from companies like Ericsson, Nokia, and others	Facilitated global connectivity, faster communication networks, and enabled innovations in IoT, autonomous vehicles, and telemedicine
Canada	Artificial Intelligence (AI) and machine learning [45, 95]	2000s–present	Pioneering work in AI research by institutions such as the University of Toronto, leading to advances in machine learning, robotics, and natural language processing	Enabled automation in various sectors like healthcare, finance, and transportation. Canada has become a leader in AI research and development, attracting global talent and investment
Germany	Automobile industry and industrial automation [103, 147]	1880s–present	Pioneering work by figures like Karl Benz and the rise of German automotive giants such as Volkswagen, BMW, and Mercedes-Benz. Germany also led in industrial automation	Germany solidified its position as the global leader in automotive manufacturing, while industrial automation advanced productivity across sectors

(continued)

Table 2 (continued)

Country	Technological advancement	Year	Key Milestones/ innovations	Impact on society and economy
Germany	Renewable energy technology [41, 114]	1990s–present	Pioneering in solar and wind energy technologies, with Germany being one of the first to implement large-scale renewable energy policies (Energiewende)	Transformation of global energy markets, reduction of carbon emissions, and promotion of sustainable economic growth in energy production
France	Nuclear energy development [60, 88]	1940s–1970s	France established itself as a leader in nuclear energy, with significant investments in nuclear power plants	Major contributor to France's energy independence, providing low-carbon energy, and positioning France as a leader in nuclear technology and exports
France	High-speed rail (TGV) [20, 36]	1981	France introduced the high-speed train (TGV), which drastically reduced travel time between major cities in France and Europe	Transformed transportation, boosting tourism and business travel, and influencing global rail systems by setting speed and efficiency standards
Italy	Automotive innovation and design [25]	1900s–present	Italy is home to renowned automotive brands like Ferrari, Fiat, and Lamborghini, known for cutting-edge design and performance engineering	Contributed to global automotive culture and luxury markets, while fostering technological advancements in car design, materials, and performance
Italy	Fashion technology [52]	1950s–present	Italy is a global leader in fashion technology, with advancements in textile production, digital fashion, and sustainable production techniques	Impacted global fashion industries, setting trends and advancing innovations in sustainable fashion and textile technology

(continued)

Table 2 (continued)

Country	Technological advancement	Year	Key Milestones/ innovations	Impact on society and economy
Japan	Robotics and automation [120]	1970s–present	Japan is a global leader in robotics, exemplified by companies like Honda, Sony, and Toyota's robotic divisions	Revolutionized manufacturing processes, healthcare, and logistics. Japan's innovations in robotics set global benchmarks for automation across industries
Japan	Consumer electronics (e.g., Walkman, PlayStation) [84]	1970s–1990s	Japan dominated the consumer electronics market with innovations such as the Walkman (Sony), and PlayStation (Sony)	Led the global entertainment and gaming revolution, influencing digital media, entertainment, and leisure industries worldwide
United Kingdom	Television and broadcasting [18]	1930s–1950s	Pioneering developments in television and broadcasting, with the BBC leading early adoption of television in the 1930s	Revolutionized global communication, entertainment, and information dissemination, shaping modern media landscapes
United Kingdom	Financial technology (fintech) [137]	2000s–present	The UK has seen explosive growth in fintech, with innovations in mobile banking, cryptocurrencies, and blockchain technologies	Transformed global financial services, enhanced access to banking and investment, and made London a key hub for fintech startups and investments
Italy	Fashion and textile innovation [89]	1950s–present	Italy is a global leader in textile technology and high-end fashion innovation, driving the adoption of digital fabrics and eco-friendly production processes	Revolutionized the global fashion industry by setting trends and innovating in textile production, promoting sustainable fashion and luxury markets
Germany	Cybersecurity and encryption [87]	2000s–present	Germany has been at the forefront of developing cybersecurity technologies, focusing on encryption and secure communications	Strengthened the global cybersecurity ecosystem, with applications ranging from governmental security to private sector data protection

Table 3 Recent developments in G7 countries

Country	Recent development	Source	Key details	Potential impacts
United States	AI regulation proposals	[96, 122]	The U.S. government has introduced new legislative proposals aimed at regulating Artificial Intelligence (AI) technologies, including deepfakes	Could lead to stricter guidelines for AI technologies, affecting tech companies and startups. May also drive innovation in compliance solutions
United Kingdom	Net zero strategy revamp	[21, 24]	The UK has updated its climate strategy to enhance its commitment to reaching net-zero emissions by 2050, with a focus on renewable energy and green jobs	Potential economic stimulus for clean tech industries. Businesses in fossil fuel sectors may face increased regulatory scrutiny and transition costs
Germany	Electric Vehicle (EV) adoption initiative	[32, 148]	Germany has launched a national initiative to incentivize electric vehicle adoption, including subsidies and infrastructure investment in charging stations	Boost for the EV market, particularly automakers and battery manufacturers. Traditional car manufacturers may need to adapt quickly to new consumer demands
France	Corporate tax reforms	[13, 50]	France has enacted tax reforms aimed at simplifying corporate taxation and incentivizing foreign investments, particularly in digital technology sectors	Positive impact on the tech and digital economy. Increased investments in French startups, but traditional sectors may see higher tax burdens
Italy	New labor market reforms	[3, 81]	Italy has implemented labor market reforms to reduce unemployment and ease restrictions on hiring and firing. Focus on youth employment and innovation sectors	Potential for increased job creation, particularly in technology and manufacturing sectors. However, challenges may arise from balancing labor rights and flexibility

(continued)

Table 3 (continued)

Country	Recent development	Source	Key details	Potential impacts
Canada	Expansion of digital infrastructure	[59, 143]	Canada is expanding its digital infrastructure with investments in broadband expansion, particularly in rural and remote areas	Improved access to digital services and education for remote communities. Increased market opportunities for digital companies in underserved regions
Japan	Work-life balance legislation	[14, 68]	Japan has passed new legislation to encourage work-life balance, limiting overtime hours and mandating paid leave for employees in tech and other sectors	There is a potential boost to employee morale and productivity, particularly in the tech and manufacturing sectors. This could influence global work culture standards

the intentional dissemination of manipulated content. For instance, while traditional defamation laws may address harmful speech or misleading statements, they struggle to account for the far-reaching implications of a doctored video or audio clip that can be disseminated globally within minutes [83, 117, 123].

From an ethical standpoint, the use of deepfakes in business raises significant questions about the authenticity of communication. The potential for these technologies to create fake corporate announcements, mimic executive voices to authorize fraudulent transactions, or manipulate consumer perception through deceptive advertising introduces an ethical dilemma about corporate transparency and truthfulness [73, 74, 132]. Companies that fail to secure their communication channels against such threats may be seen as complicit in ethical breaches, potentially losing consumer trust and credibility. Furthermore, the use of deepfakes to create fabricated endorsements or to impersonate high-ranking executives in financial decision-making poses risks not only to the affected organizations but to entire industries, disrupting market stability and investor confidence [55, 61, 97]. The ethical ramifications extend beyond fraud, as the very nature of trust in digital media is undermined, prompting significant concerns about the long-term consequences for corporate governance and business ethics in the digital era.

In the legal domain, the G7 countries face the monumental challenge of harmonizing national and international regulations to combat the growing threat of deepfake-related crimes. While intellectual property protections could theoretically shield individuals from unauthorized use of their likeness, existing frameworks struggle with the speed and scale at which deepfakes proliferate across digital platforms [9, 57, 75]. Legal scholars argue for the creation of comprehensive digital identity protection laws that would address the rights of individuals and corporations to control their likenesses and mitigate the risks posed by these synthetic forgeries. In some G7 countries, legislative bodies have proposed frameworks that criminalize

Table 4 Statistics related to G7 and deepfake

Country	Statistic/metric	Source	Details/implications	Source/study
United States	Prevalence of deepfake videos	[30, 112]	Approximately 14,000 deepfake videos were identified in 2023, a significant rise from 7,000 in 2020	Deeptrace Labs
United Kingdom	Percentage of deepfakes used for misinformation	[100]	Around 60% of deepfake videos identified in the UK were related to political or social misinformation	UK Information Commissioner's Office (ICO)
Canada	Deepfake content in online platforms	[58]	25% of all manipulated media on major online platforms (Facebook, Twitter) was identified as deepfake content	Cybersecurity Research Institute
Germany	Economic losses due to deepfake fraud	[12]	Germany lost an estimated €250 million in 2022 due to deepfake-related fraud, including business impersonations	Bundesamt für Sicherheit in der Informationstechnik (BSI)
France	Deepfake incidents in business and finance	[115, 133, 135, 136]	France reported over 100 cases of deepfake usage in corporate fraud, including fake CEO messages and financial scams	French National Cybersecurity Agency (ANSSI)
Italy	Deepfake cybersecurity threats	[134]	Over 15% of cyberattacks reported in Italy were linked to deepfake techniques used in social engineering and fraud	Italian National Cybersecurity Agency (ACN)

(continued)

Table 4 (continued)

Country	Statistic/metric	Source	Details/implications	Source/study
Japan	Incidence of deepfake sextortion cases	[109]	Japan saw a 50% increase in deepfake-based sextortion cases in 2023 compared to the previous year	National Police Agency of Japan
Canada	Research and development investment in deepfake detection	[23, 53]	Canada invested over CAD 30 million in AI and machine learning projects aimed at detecting and combating deepfakes	Government of Canada's Innovation, Science and Economic Development (ISED)

the creation and distribution of malicious deepfakes, particularly in the context of impersonation or fraud, but these efforts often remain fragmented and insufficiently robust to address the rapid evolution of the technology. The evolving nature of these legal challenges necessitates an agile, cross-jurisdictional approach to protect both individual rights and corporate interests [34, 65, 72].

Ultimately, the ethical and legal implications of deepfake technology for business practices in the G7 countries highlight the need for a concerted effort to balance innovation with regulation. As AI and machine learning technologies continue to advance, businesses must not only be vigilant in defending against these digital manipulations but also engage in proactive strategies to ensure transparency and accountability in their communications and dealings [29, 38, 43]. The evolving landscape of deepfake legislation and ethical considerations will require businesses to not only comply with new legal standards but also to embrace ethical principles that safeguard consumer trust and corporate credibility in an increasingly complex digital world.

5 Results and Discussion

The preceding analysis of deepfake technology within the G7 countries underscores both the profound capabilities and the formidable risks posed by this rapidly advancing artificial intelligence (AI) innovation. As evidenced by the literature reviewed and the statistics discussed, deepfakes represent an emerging threat to business practices that transcends traditional digital security concerns, manifesting in the domains of legal liability, corporate governance, ethical conduct, and socio-political stability [63, 118, 142]. The widespread use of generative adversarial networks (GANs) to produce hyper-realistic and nearly undetectable content has destabilized foundational notions of authenticity, creating a pressing need for more robust legal and regulatory frameworks to mitigate the risks.

The technological sophistication of deepfakes, particularly their ability to replicate human behavior with unparalleled precision, places unprecedented pressure on G7 nations to adapt their existing legal structures. Legal scholars have noted the insufficiency of current statutes to address the nuanced challenges of digital impersonation, fraud, and defamation in the context of AI-generated media. In the absence of comprehensive legislation, businesses within the G7 countries are left vulnerable to the malicious use of deepfakes, which can lead to significant financial and reputational damage [54, 66, 116]. As highlighted, incidents of impersonation, corporate fraud, and misinformation, fueled by deepfake content, have already led to tangible economic losses, such as the reported €250 million in fraud losses in Germany. Moreover, the manipulative potential of deepfakes in the corporate sector, including falsified executive communications or misleading public statements, underscores the urgent need for enhanced cybersecurity measures and proactive governance to safeguard the interests of both businesses and consumers.

The ethical implications of deepfakes for business practices are equally concerning. The ability to create fabricated endorsements, manipulate public perception, or simulate executive-level decisions introduces a host of ethical dilemmas regarding corporate transparency and truthfulness. Organizations that fail to secure themselves against deepfake exploitation risk not only immediate financial loss but also long-term erosion of public trust [2, 35, 111]. As evidenced by the increase in deepfake-related incidents across G7 countries, particularly in sectors sensitive to public image, such as finance and advertising, businesses are under growing pressure to adopt robust verification protocols to ensure the authenticity of digital communications. Failure to do so jeopardizes the ethical foundation of their operations, potentially leading to consumer backlash, reputational damage, and, in extreme cases, legal ramifications. The ethical responsibility of organizations in mitigating these risks extends beyond compliance with existing laws, necessitating a reimagining of how businesses engage with digital media in a world increasingly populated by synthetic content.

From a broader socio-economic perspective, deepfakes have profound implications for market stability, particularly in the realms of financial transactions and corporate governance. The ability to falsify corporate announcements or simulate key executives authorizing financial moves compromises the very integrity of market dynamics. Deepfake-driven market manipulation, especially in an era of high-frequency trading and global financial interconnectedness, threatens the stability of financial markets by introducing volatility and undermining investor confidence [38, 44, 82]. As evidenced by the increasing incidence of deepfake-related cyberattacks, including social engineering scams targeting both individuals and businesses, it is clear that the scale and sophistication of these threats are likely to expand, requiring urgent collaborative efforts at the international level to establish binding regulations and enforcement mechanisms.

In the context of privacy and data protection, the use of deepfakes to simulate individuals' likenesses without their consent represents a significant violation of personal privacy [15, 63, 92]. The rise of "deepfake pornography" and unauthorized use of celebrity and influencer identities for commercial purposes exacerbates these concerns, highlighting the need for stringent data protection laws that are adaptable to the challenges posed by emerging AI technologies. The ethical and legal landscape surrounding privacy rights, particularly in the G7 countries, demands an urgent reassessment of how personal data is protected, especially when AI tools make it easier to circumvent traditional privacy safeguards.

Moreover, as G7 nations grapple with these challenges, the conversation around deepfake regulation is far from homogenous. While certain countries, like the United States and the United Kingdom, have begun implementing AI regulatory frameworks, the disparity in legal approaches across jurisdictions reflects the complexity of crafting universally applicable laws in this area [86, 90, 107]. The fragmented nature of current international efforts in regulating deepfakes may leave gaps that are exploitable by malicious actors, necessitating a harmonized global approach to tackle this issue effectively. The discussion on the legal implications is thus marked by a clear divide between the need for technological innovation and the growing imperative for global consensus on regulations that can keep pace with the rapid evolution of AI [46].

In conclusion, the results of this examination demonstrate that while deepfake technology holds vast potential for innovation, its misuse within business practices in G7 countries raises significant legal, ethical, and socio-economic concerns. The interplay between AI advancements and traditional legal frameworks presents an urgent need for comprehensive regulatory measures, both at the national and international levels, to address the challenges posed by deepfakes [1, 76, 121]. As businesses continue to integrate digital technologies into their operations, it is imperative that they also fortify their cybersecurity strategies and adopt ethical business practices that prioritize authenticity and transparency [64, 94, 124]. The future of AI regulation, particularly with respect to deepfakes, will likely define the trajectory of trust in digital media and the integrity of global business practices in the years to come [42, 78, 108].

6 Conclusion

In conclusion, the advent of deepfake technology represents a paradigmatic shift in the landscape of digital content creation, one that challenges both the legal frameworks and ethical standards currently governing business practices across the G7 countries. The overwhelming capability of generative adversarial networks (GANs) to produce highly convincing and virtually indistinguishable manipulated media presents a novel set of risks that threaten the very integrity of corporate communication, public trust, and market stability. As this technology continues to evolve, its deployment—whether malicious or benign—necessitates a thorough re-evaluation of existing regulatory paradigms. The absence of clear and comprehensive laws addressing the nuances of deepfake content, coupled with the pervasive nature of these forgeries, underscores the urgent need for transnational cooperation in the formulation of governance mechanisms that can effectively curtail the proliferation of deepfakes in commercial and public domains.

From a legal perspective, the G7 nations are confronted with the challenge of adapting outdated legal doctrines to an era in which digital impersonation and identity theft have moved beyond mere textual fraud to encompass multimedia content capable of altering reality itself. The ambiguity surrounding liability—whether it rests with the creators of deepfakes, the platforms facilitating their dissemination, or the individuals exploited by these forgeries—poses a significant obstacle to effective legal intervention. This ambiguity is compounded by the rapid speed at which deepfakes can be produced and disseminated, rendering traditional legal remedies inadequate in terms of both timeliness and efficacy. Thus, there is an exigent need for international legal frameworks that not only address the creation and distribution of deepfake content but also establish clear guidelines for accountability in cases of financial manipulation, defamation, and reputational harm.

Ethically, the profound ramifications of deepfake technology on corporate conduct cannot be overstated. The potential for businesses to be hijacked by fraudulent digital representations, whether for the purpose of authorizing illicit transactions, manipulating public opinion, or engaging in deceptive advertising, challenges the very notion of ethical corporate governance. In an era where the authenticity of digital content is increasingly questioned, the imperative for companies to uphold transparency, accountability, and verifiable truth in their communications has never been more critical. Failure to do so not only exposes organizations to tangible financial losses but also imperils their long-term viability by eroding consumer trust and damaging their reputation in an increasingly distrustful digital marketplace.

Ultimately, the intersection of deepfake technology with business practices in G7 nations demands a multifaceted response that incorporates technological, legal, and ethical dimensions. While strides have been made in developing tools for detecting and mitigating deepfake content, these efforts remain insufficient in the face of a rapidly advancing technological landscape. As such, the evolution of deepfake technology serves as a clarion call for a holistic reassessment of regulatory and ethical standards governing digital media. Only through proactive, coordinated efforts across legal, technological, and ethical domains can the G7 nations hope to navigate the complexities of deepfakes, preserving both the integrity of their business practices and the trust of their constituents in the digital age.

References

1. Abdullah, K.H., Abd Aziz, F.S., Dani, R., Hammood, W.A., Setiawan, E.: Urban pollution: a bibliometric review. ASM Sci. J. **18**, 1–16 (2023). https://doi.org/10.32802/ASMSCJ.2023.1440
2. Agarwal, P., Sharma, S.: Smart urban traffic management system using energy efficient optimized path discovery. In: Proceedings of the 3rd International Conference on Artificial Intelligence and Smart Energy, ICAIS 2023, pp. 858–863 (2023). https://doi.org/10.1109/ICAIS56108.2023.10073840

3. Aggarwal, S., Sharma, S.: Voice based secured smart lock design for internet of medical things: an artificial intelligence approach. In: 2022 International Conference on Wireless Communications, Signal Processing and Networking, WiSPNET 2022, pp. 1–9 (2022). https://doi.org/10.1109/WiSPNET54241.2022.9767113
4. Al-Huqail, A.A., Kumar, P., Eid, E.M., Singh, J., Arya, A.K., Goala, M., Adelodun, B., Abou Fayssal, S., Kumar, V., Širić, I.: Risk assessment of heavy metals contamination in soil and two rice (Oryza sativa L.) varieties irrigated with paper mill effluent. Agriculture (Switzerland) **12**(11) (2022). https://doi.org/10.3390/agriculture12111864
5. AL-Huqail, A.A., Kumar, P., Eid, E.M., Taher, M.A., Kumar, P., Adelodun, B., Andabaka, Ž., Mioč, B., Držaić, V., Bachheti, A., Singh, J., Kumar, V., Širić, I.: Phytoremediation of composite industrial effluent using sacred lotus (Nelumbo nucifera Gaertn): a lab-scale experimental investigation. Sustainability (Switzerland) **14**(15) (2022). https://doi.org/10.3390/su14159500
6. Alzaidi, M.S., Subbalakshmi, C., Roshini, T.V., Shukla, P.K., Shukla, S.K., Dutta, P., Alhassan, M.: 5G-telecommunication allocation network using IoT enabled improved machine learning technique. Wirel. Commun. Mob. Comput. **2022** (2022). https://doi.org/10.1155/2022/6229356
7. Ansari, M.F., Afzal, A.: Sensitivity and performance analysis of 10 MW solar power plant using MPPT technique. Lect. Notes Electr. Eng. 894 LNEE 512–518. https://doi.org/10.1007/978-981-19-1677-9_46
8. Arora, R., Singh, A.P., Sharma, R., Chauhan, A.: A remanufacturing inventory model to control the carbon emission using cap-and-trade regulation with the hexagonal fuzzy number. Benchmarking **29**(7), 2202–2230 (2022). https://doi.org/10.1108/BIJ-05-2021-0254
9. Behera, A., Singh Rawat, K.: A brief review paper on mining subsidence and its geo-environmental impact. Mater. Today: Proceed. (2023). https://doi.org/10.1016/j.matpr.2023.04.183
10. Bhamangol, B., Kaiwade, A., Pant, B., Rana, A., Kaiwade, A., Shaikh, A.: An artificial intelligence based design and implementation for classifying the missing data in IoT applications. In: Proceedings of 5th International Conference on Contemporary Computing and Informatics, IC3I 2022, pp. 1376–1382 (2022). https://doi.org/10.1109/IC3I56241.2022.10072634
11. Bhardwaj, M., Prakash, N., Malhotra, R.K.: Energy sector and carbon trading in the future: a reference to India. In: 2024 International Conference on Smart Devices, ICSD 2024 (2024). https://doi.org/10.1109/ICSD60021.2024.10751388
12. Bhatnagar, M., Sehajpal, S.: Leveraging AI for financial sustainability in healthcare: In: Ercan, O., Azad, S., Sanjay, T., Rajendra, R., Paulo, D.J. (eds.) Sustainability, Innovation, and Consumer Preference, pp. 411–426 (2024). IGI Global Scientific Publishing. https://doi.org/10.4018/979-8-3693-9699-5.ch017
13. Bist, A., Dobriyal, R., Gwalwanshi, M., Avikal, S.: Influence of layer height and print speed on the mechanical properties of 3D-printed ABS. In: AIP Conference Proceedings, vol. 2481. American Institute of Physics Inc. (2022). https://doi.org/10.1063/5.0107304
14. Bordoloi, D., Singh, V., Sanober, S., Buhari, S.M., Ujjan, J.A., Boddu, R.: Deep learning in healthcare system for quality of service. J. Healthc. Eng. **2022**,(2022). https://doi.org/10.1155/2022/8169203
15. Caiado, R.G.G., Scavarda, L.F., Vidal, G., de Mattos Nascimento, D.L., Garza-Reyes, J.A.: A taxonomy of critical factors towards sustainable operations and supply chain management 4.0 in developing countries. Oper. Manage. Res. (2023) https://doi.org/10.1007/s12063-023-00430-8
16. Cavaliere, L.P.L., Byloppilly, R., Khan, S.D., Othman, B.A., Muda, I., Malhotra, R.K.: Acceptance and effectiveness of Industry 4.0 internal and external organisational initiatives in Malaysian firms. Int. J. Manage. Enterp. Dev. **23**(1), 1–25 (2024). https://doi.org/10.1504/IJMED.2024.138422
17. Chahar, A., Christobel, Y.A., Ritika, Adakane, P.K., Sonal, D., Tripathi, V.: The implementation of big data with cloud and edge computing in enhancing the smart grid information processes through sem model. In: 2022 2nd International Conference on Advance Computing

and Innovative Technologies in Engineering, ICACITE 2022, pp. 608–611 (2022). https://doi.org/10.1109/ICACITE53722.2022.9823896

18. Chandna, R.: Selecting the most agile manufacturing system with respect to agile attribute-technology-fuzzy AHP approach. Int. J. Oper. Res. **43**(4), 512–532 (2022). https://doi.org/10.1504/IJOR.2022.122812
19. Chandna, R., Saini, S., Kumar, S.: Selecting the most agile manufacturing system with respect to agile attribute-technology: fuzzy AHP approach. In: AIP Conference Proceedings, vol. 2481. American Institute of Physics Inc. (2022). https://doi.org/10.1063/5.0103804
20. Chandran, G.C., Synthia Regis Prabha, D.M.M., Malathi, P., Kapila, D., Arunkumar, M.S., Verma, D., Teressa, D.M.: Built-in calibration standard and decision support system for controlling structured data storage systems using soft computing techniques. Comput. Intell. Neurosci. (2022). https://doi.org/10.1155/2022/3476004
21. Chauhan, A., Sharma, N.K., Tayal, S., Kumar, V., Kumar, M.: A sustainable production model for waste management with uncertain scrap and recycled material. J. Mater. Cycles Waste Manage. **24**(5), 1797–1817 (2022). https://doi.org/10.1007/s10163-022-01435-4
22. Dani, R., Rawal, Y. S., Bagchi, P., Khan, M.: Opportunities and challenges in implementation of artificial intelligence in food & beverage service industry. In: AIP Conference Proceedings, vol. 2481. (2022). American Institute of Physics Inc. https://doi.org/10.1063/5.0103741
23. Davuluri, S.K., Alvi, S.A.M., Aeri, M., Agarwal, A., Serajuddin, M., Hasan, Z.: A security model for perceptive 5G-powered BC IoT associated deep learning. In: 6th International Conference on Inventive Computation Technologies, ICICT 2023—Proceedings, pp. 118–125 (2023). https://doi.org/10.1109/ICICT57646.2023.10134487
24. Dhiman, G., Nagar, A.K.: Editorial: blockchain-based 6G and industrial internet of things systems for Industry 4.0/5.0. Expert Syst. **39**(10) (2022). https://doi.org/10.1111/exsy.13162
25. Diddi, P.K., Sharma, P.K., Srivastava, A., Madduru, S.R.C., Reddy, E.S.: Sustainable fast setting early strength self compacting concrete (FSESSCC) using metakaolin. IOP Conf. Ser. Earth Environ. Sci. **1077**(1) (2022). https://doi.org/10.1088/1755-1315/1077/1/012009
26. Dixit, A.K., Kumar, T.V., Joshi, A., Bedi, H.S., Chakravarthi, M.K., Singh, D.P.: Trends in robotics and computer integrated manufacturing. In: Proceedings of 5th International Conference on Contemporary Computing and Informatics, IC3I 2022, pp. 212–216 (2022). https://doi.org/10.1109/IC3I56241.2022.10072745
27. Gaurav, G., Singh, A.B., Khandelwal, C., Gupta, S., Kumar, S., Meena, M.L., Dangayach, G.S.: Global development on LCA research: a bibliometric analysis from 2010 to 2021. Int. J. Soc. Ecol. Sustain. Dev. **14**(1) (2023). https://doi.org/10.4018/IJSESD.327791
28. Ghildiyal, S., Joshi, K., Rawat, G., Memoria, M., Singh, A., Gupta, A.: Industry 4.0 application in the hospitality and food service industries. In: Proceedings of the 2022 7th International Conference on Computing, Communication and Security, ICCCS 2022 and 2022 4th International Conference on Big Data and Computational Intelligence, ICBDCI 2022 (2022). https://doi.org/10.1109/ICCCS55188.2022.10079268
29. Gonfa, Y.H., Tessema, F.B., Tadesse, M.G., Bachheti, A., Bachheti, R.K.: Medicinally important plant roots and their role in nanoparticles synthesis and applications. Nanoparticles synthesis and their applications. In: Secondary Metabolites from Medicinal Plants. CRC Press (2023). https://doi.org/10.1201/9781003213727-11
30. Goswami, S., Sharma, S.: Industry 4.0 enabled molecular imaging using artificial intelligence technique. In: 2022 1st International Conference on Computational Science and Technology, ICCST 2022—Proceedings, pp. 455–460 (2022). https://doi.org/10.1109/ICCST55948.2022.10040406
31. Gupta, A., Dixit, A.K., Kumar, K.S., Lavanya, C., Chakravarthi, M.K., Gangodkar, D.: Analyzing robotics and computer integrated manufacturing of key areas using cloud computing. In: Proceedings of 5th International Conference on Contemporary Computing and Informatics, IC3I 2022, pp. 194–199 (2022). https://doi.org/10.1109/IC3I56241.2022.10072581
32. Gupta, A., Kumar, H.: Multi-dimensional perspectives on electric vehicles design: a mind map approach. Clean. Eng. Technol. **8** (2022). https://doi.org/10.1016/j.clet.2022.100483

33. Gupta, A.D., Rafi, S.M., Singh, N., Gupta, V.K., Jaiswal, S., Gangodkar, D. (2022). A framework of Internet of Things (IOT) for the manufacturing and image classifaication system. In: 2022 2nd International Conference on Advance Computing and Innovative Technologies in Engineering, ICACITE 2022, pp. 293–297 (2022). https://doi.org/10.1109/ICACITE53722.2022.9823853
34. Gupta, H., Taluja, R., Shaw, S., Chari, S.L., Deepak, A., Rana, A.: Internet of Things based reduction of electricity theft in urban areas. In: Proceedings of International Conference on Contemporary Computing and Informatics, IC3I 2023, pp. 2642–2645 (2023). https://doi.org/10.1109/IC3I59117.2023.10397868
35. Gupta, P., Gopal, S., Sharma, M., Joshi, S., Sahani, C., Ahalawat, K.: Agriculture informatics and communication: paradigm of E-governance and drone technology for crop monitoring. In: 9th International Conference on Smart Computing and Communications: Intelligent Technologies and Applications, ICSCC 2023, pp. 113–118 (2023). https://doi.org/10.1109/ICSCC59169.2023.10335058
36. Gupta, S., Kumar, V., Patil, P.: A study on recycling of waste solid garbage in a city. In: AIP Conference Proceedings, vol. 2481. American Institute of Physics Inc. (2022). https://doi.org/10.1063/5.0104563
37. Gururani, P., Bhatnagar, P., Bisht, B., Jaiswal, K.K., Kumar, V., Kumar, S., Vlaskin, M.S., Grigorenko, A.V., Rindin, K.G.: Recent advances and viability in sustainable thermochemical conversion of sludge to bio-fuel production. Fuel **316** (2022). https://doi.org/10.1016/j.fuel.2022.123351
38. Hajoary, P.K., Balachansdra, P., Garza-Reyes, J.A.: Industry 4.0 maturity and readiness assessment: an empirical validation using confirmatory composite analysis. Prod. Plan. Control. (2023). https://doi.org/10.1080/09537287.2023.2210545
39. Ishengoma, F.R., Shao, D., Alexopoulos, C., Saxena, S., Nikiforova, A.: Integration of Artificial Intelligence of Things (AIoT) in the public sector: drivers, barriers and future research agenda. Digit. Policy, Regul. Govern. **24**(5), 449–462 (2022). https://doi.org/10.1108/DPRG-06-2022-0067
40. Jena, S., Gairola, A.: Numerical method to generate and evaluate environmental wind over hills: comparison of pedestrian winds over hills and plains. CFD Lett. **14**(10), 56–67 (2022). https://doi.org/10.37934/cfdl.14.10.5667
41. Jindal, T., Sheoliha, N., Kishore, K., Uike, D., Khurana, S., Verma, D.: A conceptual analysis on the impact of Internet of Things (IOT) towards on digital marketing transformation. In: 2022 2nd International Conference on Advance Computing and Innovative Technologies in Engineering, ICACITE 2022, pp. 1943–1947 (2022). https://doi.org/10.1109/ICACITE53722.2022.9823714
42. Joshi, S., Gangola, S., Bhandari, G., Bhandari, N.S., Nainwal, D., Rani, A., Malik, S., Slama, P.: Rhizospheric bacteria: the key to sustainable heavy metal detoxification strategies. Front. Microbiol. **14**,(2023). https://doi.org/10.3389/fmicb.2023.1229828
43. Joshi, S., Sharma, M., Barve, A.: Implementation challenges of blockchain technology in closed-loop supply chain: a Waste Electrical and Electronic Equipment (WEEE) management perspective in developing countries. Supply Chain Forum **24**(1), 59–80 (2023). https://doi.org/10.1080/16258312.2022.2135972
44. Josphineleela, R., Jyothi, M., Natrayan, L., Kaviarasu, A., Sharma, M.: Development of IoT based health monitoring system for disables using microcontroller. In: Proceedings—7th International Conference on Computing Methodologies and Communication, ICCMC 2023, pp. 1380–1384 (2023). https://doi.org/10.1109/ICCMC56507.2023.10084026
45. Juneja, P.K., Kumar Sunori, S., Manu, M., Joshi, P., Sharma, S., Garia, P., Mittal, A.: Potential applications of fuzzy logic controller in the pulp and paper industry—a review. In: 5th International Conference on Inventive Computation Technologies, ICICT 2022—Proceedings, pp. 399–401 (2022). https://doi.org/10.1109/ICICT54344.2022.9850626
46. Kaliappan, S., Natrayan, L., Garg, N.: Checking and supervisory system for calculation of industrial constraints using embedded system. In: Proceedings of the 4th International Conference on Smart Electronics and Communication, ICOSEC 2023, pp. 87–90 (2023). https://doi.org/10.1109/ICOSEC58147.2023.10275952

47. Kannan, P.R., Periasamy, K., Pravin, P., Vinod Kumaar, J.R.: An experimental investigation of wire breakage and performance optimisation of WEDM process on machining of recycled aluminium alloy metal matrix composite. Mater. Sci. Poland **40**(3), 12–26 (2022). https://doi.org/10.2478/msp-2022-0030
48. Kaur, J., Khanna, R., Kumar, R., Sunil, G.: Role of blockchain technologies in goods and services tax. In: Proceedings—2024 3rd International Conference on Sentiment Analysis and Deep Learning, ICSADL 2024, pp. 607–612 (2024). https://doi.org/10.1109/ICSADL61749.2024.00104
49. Khanna, R., Sharma, S., Czeh, T., Adhana, K.: Challenges, interventions, and strategies for higher education transformation: vision 2047. Lect. Notes Netw. Syst. 1075 LNNS, 637–655 (2025). https://doi.org/10.1007/978-981-97-6106-7_38
50. Khaudiyal, S., Rawat, A., Das, S.K., Garg, N.: Bacterial concrete: a review on self-healing properties in the light of sustainability. Mater. Today: Proceed. **60**, 136–143 (2022). https://doi.org/10.1016/j.matpr.2021.12.277
51. Kohli, P., Sharma, S., Matta, P.: Secured authentication schemes of 6G driven vehicular communication network in Industry 5.0 Internet-of-Everything (IoE) applications: challenges and opportunities. In: 2022 IEEE 2nd International Conference on Mobile Networks and Wireless Communications, ICMNWC 2022 (2022a). https://doi.org/10.1109/ICMNWC56175.2022.10031781
52. Kohli, P., Sharma, S., Matta, P.: Secured privacy preserving techniques analysis of 6G driven vehicular communication network in Industry 5.0 Internet-of-Everything (IoE) applications. In: 2022 International Conference on Smart Generation Computing, Communication and Networking, SMART GENCON 2022 (2022b). https://doi.org/10.1109/SMARTGENCON56628.2022.10084289
53. Komkowski, T., Antony, J., Garza-Reyes, J.A., Tortorella, G.L., Pongboonchai-Empl, T.: A systematic review of the integration of Industry 4.0 with quality-related operational excellence methodologies. Qual. Manage. J. **30**(1), 3–15 (2023a). https://doi.org/10.1080/10686967.2022.2144783
54. Komkowski, T., Antony, J., Garza-Reyes, J.A., Tortorella, G.L., Pongboonchai-Empl, T.: Integrating lean management with Industry 4.0: an explorative dynamic capabilities theory perspective. In: Production Planning and Control (2023b). https://doi.org/10.1080/09537287.2023.2294297
55. Kowsalya, K., Rani, R.P.J., Ritu, Bhiyana, M., Saini, M., Patil, P.P.: Blockchain-internet of things-machine learning: development of traceable system for multi purposes. In: 2023 3rd International Conference on Advance Computing and Innovative Technologies in Engineering, ICACITE 2023, pp. 1112–1115 (2023). https://doi.org/10.1109/ICACITE57410.2023.10183065
56. Krishna, S.H., Upadhyay, A., Tewari, M., Gehlot, A., Girimurugan, B., Pundir, S.: Empirical investigation of the key machine learning elements promoting e-business using an SEM framework. In: Proceedings of 5th International Conference on Contemporary Computing and Informatics, IC3I 2022, pp. 1960–1964 (2022). https://doi.org/10.1109/IC3I56241.2022.10072712
57. Kukreti, A., Shriyal, A., Sharma, S., Bhadula, S.: Internet-of-Things enabled smart and portable terrace garden protection shed. In: 2023 4th IEEE Global Conference for Advancement in Technology, GCAT 2023 (2023). https://doi.org/10.1109/GCAT59970.2023.10353281
58. Kukreti, M., Sehajpal, S., Tiwari, R., Sood, K.: Driving Global Health and Sustainable Development Goals With Smart Technology. In: Mohit, K., Sabina, S., Rajesh, T., Kiran, S. (eds.), 1st ed. IGI Global Scientific Publishing. https://doi.org/10.4018/979-8-3373-0240-9
59. Kumar, K.S., Yadav, D., Joshi, S.K., Chakravarthi, M.K., Jain, A.K., Tripathi, V.: Blockchain technology with applications to distributed control and cooperative robotics. In: Proceedings of 5th International Conference on Contemporary Computing and Informatics, IC3I 2022, pp. 206–211 (2022). https://doi.org/10.1109/IC3I56241.2022.10073275

60. Kumar, M.A., Prasad, M.S.G., More, P., Christa, S.: Artificial intelligence-based personal health monitoring devices. In: Mobile Health: Advances in Research and Applications—Volume II. Nova Science Publishers, Inc. (2022)
61. Kumar, N.M., Islam, S., Podder, A.K., Selim, A., Bajaj, M., Kamel, S.: Lifecycle-based feasibility indicators for floating solar photovoltaic plants along with implementable energy enhancement strategies and framework-driven assessment approaches leading to advancements in the simulation tool. Front. Energy Res. **11** (2023). https://doi.org/10.3389/fenrg.2023.1075384
62. Kumar, R., Goel, R., Singh, T., Mohanty, S.M., Gupta, D., Alkhayyat, A., Khanna, R.: Sustainable finance factors in indian economy: analysis on policy of climate change and energy sector. Fluctuation Noise Lett. **23**(2) (2024). https://doi.org/10.1142/S0219477524400042
63. Kumar, R., Kandpal, B., Ahmad, V.: Industrial IoT (IIOT): security threats and countermeasures. In: International Conference on Innovative Data Communication Technologies and Application, ICIDCA 2023—Proceedings, pp. 829–833 (2023). https://doi.org/10.1109/ICIDCA56705.2023.10100145
64. Kumar, R., Khanna, R.: RPA (Robotic Process Automation) in finance & accounting and future scope. In: Proceedings of the 2023 2nd International Conference on Augmented Intelligence and Sustainable Systems, ICAISS 2023, pp. 1640–1645 (2023). https://doi.org/10.1109/ICAISS58487.2023.10250496
65. Kumar, R., Saxena, A., Singh, R.: Robotic Process automation bridge-in banking institute and consumers. In: 2023 International Conference on Disruptive Technologies, ICDT 2023, pp. 428–431 (2023). https://doi.org/10.1109/ICDT57929.2023.10150500
66. Lade, J., Mohammed, K.A., Singh, D., Prasad Verma, R., Math, P., Saraswat, M., Raj Gupta, L.: A critical review of fabrication routes and their effects on mechanical properties of AMMCs. Mater. Today: Proceed. (2023). https://doi.org/10.1016/j.matpr.2023.03.041
67. Lokanadham, D., Sharma, R.C., Palli, S., Bhardawaj, S.: Wear rate modelling and analysis of limestone slurry particulate composites using the fuzzy method. Int. J. Recent Innov. Trends Comput. Commun. **10**(1), 133–143 (2022). https://doi.org/10.17762/ijritcc.v10i1s.5818
68. Lourens, M., Tamizhselvi, A., Goswami, B., Alanya-Beltran, J., Aarif, M., Gangodkar, D.: Database management difficulties in the Internet of Things. In: Proceedings of 5th International Conference on Contemporary Computing and Informatics, IC3I 2022, pp. 322–326 (2022). https://doi.org/10.1109/IC3I56241.2022.10072614
69. Malhotra, R.K., Sharma, C.A.P., Upreti, H., Sharma, S., Prakash, N., Mittal, S.: Role of psychological contract in employee engagement in higher education: empirical study on Uttarakhand. In: E3S Web of Conferences, vol. 556. EDP Sciences (2024). https://doi.org/10.1051/e3sconf/202455601030
70. Mandalapu, S.R., Sivamuni, K., Chitra Devi, D., Aswal, U.S., Sherly, S.I., Balaji, N.A.: An architecture-based self-typing service for cloud native applications. In: Proceedings of the 4th International Conference on Smart Electronics and Communication, ICOSEC 2023, pp. 562–566 (2023). https://doi.org/10.1109/ICOSEC58147.2023.10276313
71. Masal, V., Pavithra, P., Tiwari, S.K., Singh, R., Panduro-Ramirez, J., Gangodkar, D.: Deep learning applications for blockchain in industrial IoT. In: Proceedings of 5th International Conference on Contemporary Computing and Informatics, IC3I 2022, pp. 276–281 (2022). https://doi.org/10.1109/IC3I56241.2022.10073357
72. Medhi, M.K., Ambust, S., Kumar, R., Das, A.J.: Characterization and purification of biosurfactants. In: Advancements in Biosurfactants Research. Springer International Publishing (2023). https://doi.org/10.1007/978-3-031-21682-4_4
73. Mehershilpa, G., Prasad, D., Sai Kiran, C., Shaikh, A., Jayashree, K., Socrates, S.: EDM machining of Ti6Al4V alloy using colloidal biosilica. Mater. Today: Proceed. (2023) https://doi.org/10.1016/j.matpr.2023.02.443
74. Mishra, A.K., Wazid, M.: Design of a cloud-based security mechanism for Industry 4.0 communication. In: ICSCCC 2023—3rd International Conference on Secure Cyber Computing and Communications, pp. 337–343 (2023). https://doi.org/10.1109/ICSCCC58608.2023.10176702

75. Mohamed, N., Sridhara Rao, L., Sharma, M., Sureshbaburajasekaranl, Badriasulaimanal-furhood, Kumar Shukla, S.: In-depth review of integration of AI in cloud computing. In: 2023 3rd International Conference on Advance Computing and Innovative Technologies in Engineering, ICACITE 2023, pp. 1431–1434 (2023). https://doi.org/10.1109/ICACITE57410.2023.10182738
76. Mohd, N., Kumar, I., Khurshid, A.A.: Changing roles of intelligent robotics and machinery control systems as Cyber-Physical Systems (CPS) in the Industry 4.0 framework. In: 2023 International Conference on Communication, Security and Artificial Intelligence, ICCSAI 2023, pp. 647–651 (2023). https://doi.org/10.1109/ICCSAI59793.2023.10421085
77. Nainwal, P., Lall, S., Nawaz, A.: Physiochemical characterization of silver nanoparticles using rhizome extract of Alpinia galanga and its antimicrobial activity. J. Med. Pharm. Allied Sci. Int. Conf. (2), 219–223 (2022). https://doi.org/10.22270/jmpas.VIC2I2.1830
78. Naithani, D., Khandelwal, R.R., Garg, N.: Development of an automobile hardware-inthe-loop test system with CAN communication. In: Proceedings of the 2023 2nd International Conference on Augmented Intelligence and Sustainable Systems, ICAISS 2023, pp. 1653–1656 (2023). https://doi.org/10.1109/ICAISS58487.2023.10250529
79. Pai, H.A., Almuzaini, K.K., Ali, L., Javeed, A., Pant, B., Pareek, P.K., Akwafo, R.: Delay-driven opportunistic routing with multichannel cooperative neighbor discovery for Industry 4.0 wireless networks based on power and load awareness. Wirel. Commun. Mobile Comput. (2022). https://doi.org/10.1155/2022/5256133
80. Pallavi, B., Othman, B., Trivedi, G., Manan, N., Pawar, R.S., Singh, D.P.: The application of the Internet of Things (IoT) to establish a technologically advanced Industry 4.0 for long-term growth and development. In: 2022 2nd International Conference on Advance Computing and Innovative Technologies in Engineering, ICACITE 2022, pp. 1927–1932 (2022). https://doi.org/10.1109/ICACITE53722.2022.9823481
81. Pandey, K., Paliwal, S., Joshi, H., Bisht, N., Kumar, N.: A review on change control: a critical process of the pharmaceutical industry. J. Med. Pharm. Allied Sci. **11**(2), 4588–4592 (2022). https://doi.org/10.55522/jmpas.V11I2.2077
82. Pandey, N.K., Kashyap, S., Sharma, A., Diwakar, M.: Contribution of cloud-based services in post-pandemic technology sustainability and challenges: a future direction. In: Evolving Networking Technologies: Developments and Future Directions. Wiley (2023)
83. Pandya, D.J., Kumar, Y., Singh, D.P., Vairavel, D.K., Deepak, A., Rao, A.K., Rana, A.: Automatic power factor compensation for industrial use to minimize penalty. In: Proceedings of International Conference on Contemporary Computing and Informatics, IC3I 2023, pp. 2499–2504. https://doi.org/10.1109/IC3I59117.2023.10398095
84. Praveenchandar, J., Vetrithangam, D., Kaliappan, S., Karthick, M., Pegada, N.K., Patil, P.P., Rao, S.G., Umar, S.: IoT-based harmful toxic gases monitoring and fault detection on the sensor dataset using deep learning techniques. Sci. Program. **2022** (2022). https://doi.org/10.1155/2022/7516328
85. Rajawat, A.S., Singh, S., Gangil, B., Ranakoti, L., Sharma, S., Asyraf, M.R.M., Razman, M.R.: Effect of marble dust on the mechanical, morphological, and wear performance of basalt fibre-reinforced epoxy composites for structural applications. Polymers **14**(7) (2022). https://doi.org/10.3390/polym14071325
86. Rajbalaji, S., Raman, R., Pant, B., Rathour, N., Rajagopa, B.R., Prasad, C.R.: Design of deep learning models for the identifications of harmful attack activities in IIOT. In: 2023 International Conference on Artificial Intelligence and Smart Communication, AISC 2023, pp. 609–613 (2023). https://doi.org/10.1109/AISC56616.2023.10085088
87. Rajeswari, M., Kumar, N., Raman, P., Patjoshi, P.K., Singh, V., Pundir, S.: Optimal analysis for enterprise financial management based on artificial intelligence and parallel computing method. In: Proceedings of 5th International Conference on Contemporary Computing and Informatics, IC3I 2022, pp. 2081–2086 (2022). https://doi.org/10.1109/IC3I56241.2022.10072851
88. Raju, K., Balakrishnan, M., Prasad, D.V.S.S.S.V., Nagalakshmi, V., Patil, P.P., Kaliappan, S., Arulmurugan, B., Radhakrishnan, K., Velusamy, B., Paramasivam, P., El-Denglawey, A.:

Optimization of WEDM process parameters in Al2024-Li-Si3N4MMC. J. Nanomater. **2022** (2022). https://doi.org/10.1155/2022/2903385
89. Ram, M., Bhandari, A.S., Kumar, A.: Reliability evaluation and cost optimization of solar road studs. Int. J. Reliab. Qual. Saf. Eng. **29**(1) (2022). https://doi.org/10.1142/S0218539321500418
90. Ram, M., Xing, L.: Reliability modeling in industry 4.0. In: Reliability Modeling in Industry 4.0. Elsevier (2023). https://doi.org/10.1016/C2021-0-01679-5
91. Ramakrishnan, T., Mohan Gift, M.D., Chitradevi, S., Jegan, R., Subha Hency Jose, P., Nagaraja, H.N., Sharma, R., Selvakumar, P., Hailegiorgis, S.M.: Study of numerous resins used in polymer matrix composite materials. Adv. Mater. Sci. Eng. (2022). https://doi.org/10.1155/2022/1088926
92. Raman, R., Kumar, R., Ghai, S., Gehlot, A., Raju, A.M., Barve, A.: A new method of optical spectrum analysis for advanced wireless communications. In: 2023 3rd International Conference on Advance Computing and Innovative Technologies in Engineering, ICACITE 2023, pp. 1719–1723 (2023). https://doi.org/10.1109/ICACITE57410.2023.10182414
93. Rana, M.S., Cavaliere, L.P.L., Mishra, A.B., Padhye, P., Singh, R.R., Khanna, R.: Internet of Things (IOT) based assessment for effective monitoring data against malicious attacks on financial collectors. In: 2022 2nd International Conference on Advance Computing and Innovative Technologies in Engineering, ICACITE 2022, pp. 177–181 (2022). https://doi.org/10.1109/ICACITE53722.2022.9823612
94. Rawat, R., Sharma, S., Goyal, H.R.: Intelligent digital financial inclusion system architectures for industry 5.0 enabled digital society. Winter Summit Smart Comput. Netw. WiSSCoN 2023. https://doi.org/10.1109/WiSSCoN56857.2023.10133858
95. Rawat, S.S., Pant, S., Kumar, A., Ram, M., Sharma, H.K., Kumar, A.: A state-of-the-art survey on analytical hierarchy process applications in sustainable development. Int. J. Math. Eng. Manage. Sci. **7**(6), 883–917. https://doi.org/10.33889/IJMEMS.2022.7.6.056
96. Saini, S., Sachdeva, L., Badhotiya, G.K.: Sustainable human resource management: a conceptual framework. ECS Trans. **107**(1), 6455–6463 (2022). https://doi.org/10.1149/10701.6455ecst
97. Sathyaseelan, K., Vyas, T., Madala, R., Chamundeeswari, V., Rai Goyal, H., Jayaraman, R.: Blockchain enabled intelligent surveillance system model with AI and IoT. In: Proceedings of 8th IEEE International Conference on Science, Technology, Engineering and Mathematics, ICONSTEM 2023 (2023). https://doi.org/10.1109/ICONSTEM56934.2023.10142303
98. Sati, P., Sharma, E., Soni, R., Dhyani, P., Solanki, A.C., Solanki, M.K., Rai, S., Malviya, M.K.: Bacterial endophytes as bioinoculant: microbial functions and applications toward sustainable farming. In: Microbial Endophytes and Plant Growth: Beneficial Interactions and Applications. Elsevier (2022). https://doi.org/10.1016/B978-0-323-90620-3.00008-8
99. Saxena, A., Pant, B., Alanya-Beltran, J., Akram, S.V., Bhaskar, B., Bansal, R.: A detailed review of implementation of deep learning approaches for industrial internet of things with the different opportunities and challenges. In: Proceedings of 5th International Conference on Contemporary Computing and Informatics, IC3I 2022, pp. 1370–1375 (2022). https://doi.org/10.1109/IC3I56241.2022.10072499
100. Sehajpal, S., Kaur, N.: Neuroscience, nursing, and financial innovation: a synergistic approach to optimized patient care and economic well-being. In: Sanjay, T., Bhupinderpal Pal, S.C., Amar, J., Ercan, O., Pawan, K. (eds.) Neuromarketing's Role in Sustainable Finance, pp. 133–146. IGI Global Scientific Publishing (2025). https://doi.org/10.4018/979-8-3693-9117-4.ch008
101. Shabbiruddin, Kanwar, N., Jadoun, V.K., Jayalakshmi, N.S.J., Afthanorhan, A., Fatema, N., Malik, H., Hossaini, M.A.: Industry—challenge to pro-environmental manufacturing of goods replacing single-use plastic by Indian industry: a study toward failing ban on single-use plastic access. IEEE Access **11**, 77336–77346 (2023). https://doi.org/10.1109/ACCESS.2023.3296097
102. Sharma, K., Pandit, S., Thapa, B.S., Pant, M.: Biodegradation of congo red using co-culture anode inoculum in a microbial fuel cell. Catalysts **12**(10) (2022). https://doi.org/10.3390/catal12101219

103. Sharma, M., Hagar, A.A., Krishna Murthy, G.R., Beyane, K., Gawali, B.W., Pant, B.: A study on recognising the application of multiple big data technologies and its related issues, difficulties and opportunities. In: 2022 2nd International Conference on Advance Computing and Innovative Technologies in Engineering, ICACITE 2022, pp. 341–344 (2022). https://doi.org/10.1109/ICACITE53722.2022.9823623
104. Sharma, M., Luthra, S., Joshi, S., Joshi, H.: Challenges to agile project management during COVID-19 pandemic: an emerging economy perspective. Oper. Manag. Res. **15**(1–2), 461–474 (2022). https://doi.org/10.1007/s12063-021-00249-1
105. Sharma, P., Malhotra, R.K., Altybekova, D., Singh, C.P.: AI and machine learning as catalysts for educational equity: bridging the urban–rural divide. Lect. Notes Netw. Syst. 1074 LNNS 41–55 (2024). https://doi.org/10.1007/978-981-97-6103-6_4
106. Sharma, P., Malhotra, R.K., Kumar, R.: How is digitalization reshaping the landscape of rural banking in India? In: 2024 International Conference on Smart Devices, ICSD 2024 (2024). https://doi.org/10.1109/ICSD60021.2024.10750981
107. Sharma, R.C., Palli, S., Sharma, S.K.: Ride analysis of railway vehicle considering rigidity and flexibility of the carbody. J. Chin. Inst. Eng. Trans. Chin. Inst. Eng. Ser. A **46**(4), 355–366 (2023). https://doi.org/10.1080/02533839.2023.2194918
108. Sharma, S., Bhadula, S.: Secure federated learning for intelligent industry 4.0 IoT enabled self skin care application system. In: Proceedings of the 2nd International Conference on Applied Artificial Intelligence and Computing, ICAAIC 2023, pp. 1164–1170 (2023). https://doi.org/10.1109/ICAAIC56838.2023.10141028
109. Sharma, S., Gupta, A., Tyagi, R.: sustainable natural resources utilization decision system for better society using vedic scripture, cloud computing, and IoT. In: Proceedings of IEEE 2023 5th International Conference on Advances in Electronics, Computers and Communications, ICAECC 2023. Institute of Electrical and Electronics Engineers Inc. (2023). https://doi.org/10.1109/ICAECC59324.2023.10560335
110. Sharma, S., Gupta, C., Khanna Malhotra, R., Upreti, H.: Sustainable banking practices: impact, challenges and opportunities. In: E3S Web of Conferences, vol. 556. EDP Sciences (2024). https://doi.org/10.1051/e3sconf/202455601031
111. Sharma, S., Kandpal, V., Choudhury, T., Santibanez Gonzalez, E.D.R., Agarwal, N.: Assessment of the implications of energy-efficient technologies on the environmental sustainability of rail operation. AIMS Environ. Sci. **10**(5), 709–731 (2023). https://doi.org/10.3934/environsci.2023039
112. Shashikala, R., Singh, B.P., Azam, M., Magesh, C.R., Rajat, Singh, D.P.: IoT engineering nanomaterial's approach to sustainable advance crop production management. In: 2022 2nd International Conference on Advance Computing and Innovative Technologies in Engineering, ICACITE 2022, pp. 2284–2288 (2022). https://doi.org/10.1109/ICACITE53722.2022.9823573
113. Shukla, S.K., Pant, B., Viriyasitavat, W., Verma, D., Kautish, S., Dhiman, G., Kaur, A., Srihari, K., Mohanty, S.N.: An integration of autonomic computing with multicore systems for performance optimization in Industrial Internet of Things. IET Commun. (2022). https://doi.org/10.1049/cmu2.12505
114. Singamaneni, K.K., Dhiman, G., Juneja, S., Muhammad, G., AlQahtani, S.A., Zaki, J.: A novel QKD approach to enhance IIOT privacy and computational knacks. Sensors **22**(18) (2022). https://doi.org/10.3390/s22186741
115. Singh, A., Kumar, P., Taneja, S.: Artificial intelligence and machine learning-powered smart finance. In: Sanjay Taneja, P.K., Singh, A. (eds.) IGI Global. IGI Global (2024). https://doi.org/10.4018/979-8-3693-3264-1
116. Singh, K.D., Deep Singh, P., Bansal, A., Kaur, G., Khullar, V., Tripathi, V.: Exploratory data analysis and customer churn prediction for the telecommunication industry. In: ACCESS 2023–2023 3rd International Conference on Advances in Computing, Communication, Embedded and Secure Systems, pp. 197–201 (2023). https://doi.org/10.1109/ACCESS57397.2023.10199700

117. Singh, K.D., Singh, P., Chhabra, R., Kaur, G., Bansal, A., Tripathi, V.: Cyber-physical systems for smart city applications: a comparative study. In: 2023 International Conference on Advancement in Computation and Computer Technologies, InCACCT 2023, pp. 871–876. Institute of Electrical and Electronics Engineers Inc. (2023). https://doi.org/10.1109/InCACCT57535.2023.10141719
118. Singh, K.D., Singh, P., Kaur, G., Khullar, V., Chhabra, R., Tripathi, V.: Education 4.0: exploring the potential of disruptive technologies in transforming learning. In: Proceedings of International Conference on Computational Intelligence and Sustainable Engineering Solution, CISES 2023, pp. 586–591 (2023). https://doi.org/10.1109/CISES58720.2023.10183547
119. Singh, K.D., Singh, P., Tripathi, V., Khullar, V.: A novel and secure framework to detect unauthorized access to an optical fog-cloud computing network. In: PDGC 2022–2022 7th International Conference on Parallel, Distributed and Grid Computing, pp. 618–622. Institute of Electrical and Electronics Engineers Inc. (2022). https://doi.org/10.1109/PDGC56933.2022.10053223
120. Singh, R., Chandra, A.S., Bbhagat, B., Panduro-Ramirez, J., Gaikwad, A.P., Pant, B.: Cloud computing, machine learning, and secure data sharing enabled through blockchain. In: Proceedings of 5th International Conference on Contemporary Computing and Informatics, IC3I 2022, pp. 282–286 (2022). https://doi.org/10.1109/IC3I56241.2022.10072925
121. Singh, S.K., Chauhan, A., Sarkar, B.: Sustainable biodiesel supply chain model based on waste animal fat with subsidy and advertisement. J. Clean. Prod. **382** (2023). https://doi.org/10.1016/j.jclepro.2022.134806
122. Singh, Y., Rahim, E.A., Singh, N.K., Sharma, A., Singla, A., Palamanit, A.: Friction and wear characteristics of chemically modified mahua (madhuca indica) oil based lubricant with SiO_2 nanoparticles as additives. Wear **508–509** (2022). https://doi.org/10.1016/j.wear.2022.204463
123. Singla, A., Singh, Y., Singh, Y., Rahim, E.A., Singh, N.K., Sharma, A.: Challenges and future prospects of biofuel generations: an overview. In: Biowaste and Biomass in Biofuel Applications. CRC Press (2023). https://doi.org/10.1201/9781003265597-4
124. Singla, R.K., De, R., Efferth, T., Mezzetti, B., Sahab Uddin, M., Sanusi, Ntie-Kang, F., Wang, D., Schultz, F., Kharat, K.R., Devkota, H.P., Battino, M., Sur, D., Lordan, R., Patnaik, S.S., Tsagkaris, C., Sai, C.S., Tripathi, S.K., Găman, M.-A., Shen, B.: The International Natural Product Sciences Taskforce (INPST) and the power of Twitter networking exemplified through #INPST hashtag analysis. Phytomedicine **108** (2023). https://doi.org/10.1016/j.phymed.2022.154520
125. Širić, I., Eid, E.M., Taher, M.A., El-Morsy, M.H.E., Osman, H.E.M., Kumar, P., Adelodun, B., Abou Fayssal, S., Mioč, B., Andabaka, Ž., Goala, M., Kumari, S., Bachheti, A., Choi, K.S., Kumar, V.: Combined use of spent mushroom substrate biochar and PGPR improves growth, yield, and biochemical response of cauliflower (Brassica oleracea var. botrytis): a preliminary study on greenhouse cultivation. Horticulturae **8**(9) (2022). https://doi.org/10.3390/horticulturae8090830
126. Sonnad, S., Awasthy, M., Rane, K., Banerjee, M., Buddhi, D., Pant, B.: Blockchain-based secure mengers authentication for industrial IoT. In: Proceedings of the 2022 11th International Conference on System Modeling and Advancement in Research Trends, SMART 2022, pp. 853–858. Institute of Electrical and Electronics Engineers Inc. (2022). https://doi.org/10.1109/SMART55829.2022.10046934
127. Srivastava, A., Jawaid, S., Singh, R., Gehlot, A., Akram, S.V., Priyadarshi, N., Khan, B.: Imperative role of technology intervention and implementation for automation in the construction industry. Adv. Civ. Eng. **2022** (2022). https://doi.org/10.1155/2022/6716987
128. Srivastava, S.P., Malhotra, R.K., Mukherjee, R.K.: Data driven decision making in insurance. In: 2024 International Conference on Smart Devices, ICSD 2024 (2024). https://doi.org/10.1109/ICSD60021.2024.10751068
129. Subramani, R., Kaliappan, S., Kumar, P.V.A., Sekar, S., De Poures, M.V., Patil, P.P., Raj, E.S.E.: A recent trend on additive manufacturing sustainability with supply chain management

concept, multicriteria decision making techniques. Adv. Mater. Sci. Eng. **2022** (2022). https://doi.org/10.1155/2022/9151839
130. Subramani, R., Kaliappan, S., Sekar, S., Patil, P.P., Usha, R., Manasa, N., Esakkiraj, E.S.: Polymer filament process parameter optimization with mechanical test and morphology analysis. Adv. Mater. Sci. Eng. **2022** (2022). https://doi.org/10.1155/2022/8259804
131. Syed, F.A., Bargavi, N., Sharma, A., Mishra, A., Nagpal, P., Srivastava, A.: Recent management trends involved with the Internet of Things in Indian automotive components manufacturing industries. In: Proceedings of 5th International Conference on Contemporary Computing and Informatics, IC3I 2022, pp. 1035–1041 (2022). https://doi.org/10.1109/IC3I56241.2022.10072565
132. Tamilmani, S., Mohan, T., Jeyalakshmi, S., Shukla, G.P., Gehlot, A., Shukla, S.K.: Blockchain integrated with industrial IOT towards industry 4.0. In: 2023 International Conference on Artificial Intelligence and Smart Communication, AISC 2023, pp. 575–581 (2023). https://doi.org/10.1109/AISC56616.2023.10085226
133. Taneja, S., Chahal, B.P.S., Johri, A., Ozen, E., Kumar, P.: Neuromarketing's Role in Sustainable Finance. IGI Global (2024). https://doi.org/10.4018/979-8-3693-9117-4
134. Taneja, S., Gupta, S., Singh, A., Bhatnagar, M., Gaur, A.: Green management—a new paradigm in the world of business. Green Manage. A New Paradigm World Bus. 1–230 (2023). https://doi.org/10.52305/DLIH6622
135. Taneja, S., Kumar, P., Reepu, Kukreti, M., Ozen, E.: (2024) Data Alchemy in the Insurance Industry. Emerald Publishing Limited (2024). https://doi.org/10.1108/978-1-83608-582-920241033
136. Taneja, S., Kumar, P., Reepu, Ozen, E.: E-finance: transformative strategies for sustainable development in the digital era. In: E-Finance: Transformative Strategies for Sustainable Development in the Digital Era. Nova Science Publishers, Inc. (2025). https://doi.org/10.52305/NZSJ1375
137. Thapa, B. Sen, Pandit, S., Patwardhan, S. B., Tripathi, S., Mathuriya, A.S., Gupta, P.K., Lal, R.B., Tusher, T.R.: Application of Microbial Fuel Cell (MFC) for pharmaceutical wastewater treatment: an overview and future perspectives. Sustainability (Switzerland) **14**(14) (2022). https://doi.org/10.3390/su14148379
138. Thentral, T.M.T., Usha, S., Palanisamy, R., Geetha, A., Alkhudaydi, A.M., Sharma, N.K., Bajaj, M., Ghoneim, S.S.M., Shouran, M., Kamel, S.: An energy efficient modified passive power filter for power quality enhancement in electric drives. Front. Energy Res. **10** (2022). https://doi.org/10.3389/fenrg.2022.989857
139. Trache, D., Tarchoun, A.F., Abdelaziz, A., Bessa, W., Hussin, M.H., Brosse, N., Thakur, V.K.: Cellulose nanofibrils-graphene hybrids: recent advances in fabrication, properties, and applications. Nanoscale **14**(35), 12515–12546 (2022). https://doi.org/10.1039/d2nr01967a
140. Tyagi, S., Jindal, T., Krishna, S.H., Hassen, S.M., Shukla, S.K., Kaur, C.: Comparative analysis of artificial intelligence and its powered technologies applications in the finance sector. In: Proceedings of 5th International Conference on Contemporary Computing and Informatics, IC3I 2022, pp. 260–264 (2022). https://doi.org/10.1109/IC3I56241.2022.10073077
141. Uike, D., Agarwalla, S., Bansal, V., Chakravarthi, M.K., Singh, R., Singh, P.: Investigating the role of block chain to secure identity in IoT for industrial automation. In: Proceedings of the 2022 11th International Conference on System Modeling and Advancement in Research Trends, SMART 2022, pp. 837–841. Institute of Electrical and Electronics Engineers Inc. (2022). https://doi.org/10.1109/SMART55829.2022.10047385
142. Umamaheswaran, S.K., Singh, G., Dixit, A.K., Mc, S.C., Chakravarthi, M.K., Singh, D.P.: IOT-based analysis for effective continuous monitoring prevent fraudulent intrusions in finance and banking. In: 2023 International Conference on Artificial Intelligence and Smart Communication, AISC 2023, pp. 548–552 (2023). https://doi.org/10.1109/AISC56616.2023.10084920
143. Unhelkar, B., Joshi, S., Sharma, M., Prakash, S., Mani, A.K., Prasad, M.: Enhancing supply chain performance using RFID technology and decision support systems in the industry 4.0—a systematic literature review. Int. J. Inf. Manage. Data Insights **2**(2). https://doi.org/10.1016/j.jjimei.2022.100084

144. Uniyal, S., Sarma, P.R.S., Kumar Mangla, S., Tseng, M.-L., Patil, P.: ICT as "knowledge management" for assessing sustainable consumption and production in supply chains. In: Research Anthology on Measuring and Achieving Sustainable Development Goals, vol. 3. IGI Global. https://doi.org/10.4018/978-1-6684-3885-5.ch048
145. Upreti, H., Malhotra, R.K.: Bridging the urban-rural education gap in India through CSR (Corporate Social Responsibility) initiatives: a conceptual study with special reference to sustainable development goal 4 (Quality Education). In: E3S Web of Conferences, vol. 556 (2024). EDP Sciences. https://doi.org/10.1051/e3sconf/202455601032
146. Upreti, H., Malhotra, R. K., Kumar, R., Karki, B.: leveraging smart devices for corporate social responsibility in India: advancements, trends, and concerns towards sustainability. In: 2024 International Conference on Smart Devices, ICSD 2024 (2024). https://doi.org/10.1109/ICSD60021.2024.10751578
147. Verma, M., Ahmad, W., Park, J.-H., Kumar, V., Vlaskin, M.S., Vaya, D., Kim, H.: One-step functionalization of chitosan using EDTA: kinetics and isotherms modeling for multiple heavy metals adsorption and their mechanism. J. Water Process Eng. **49** (2022). https://doi.org/10.1016/j.jwpe.2022.102989
148. Verma, P., Chaudhari, V., Dumka, A., Singh, R.P.: A meta-analytical review of deep learning prediction models for big data. In: Encyclopedia of Data Science and Machine Learning. IGI Global. https://doi.org/10.4018/978-1-7998-9220-5.ch023
149. Verma, P., Kumar, V., Daim, T., Sharma, N.K., Mittal, A.: Identifying and prioritizing impediments of industry 4.0 to sustainable digital manufacturing: a mixed method approach. J. Clean. Prod. 356 (2022). https://doi.org/10.1016/j.jclepro.2022.131639
150. Yeruva, A.R., Vijaya Durga, C.S.L., Gokulavasan, B., Pant, K., Chaturvedi, P., Srivastava, A.P.: A smart healthcare monitoring system based on fog computing architecture. In: Proceedings of International Conference on Technological Advancements in Computational Sciences, ICTACS 2022, pp. 904–909. https://doi.org/10.1109/ICTACS56270.2022.9987881

Navigating the Digital Mirage: A Comparative Analysis of Deepfake Challenges and Responses Across BRICS Nations and the Global Arena

Kavita Dahiya and Umang Sharma

Abstract The rapid proliferation of deepfake technology has ushered in a new era of digital manipulation, presenting significant challenges and ethical quandaries for global governance and organizational integrity. This research examines the positioning of BRICS nations—Brazil, Russia, India, China, and South Africa—in the context of deepfake regulation, technological advancement, and public awareness, drawing comparative insights with the broader global response to this emergent phenomenon. Through a meticulous analysis of legal frameworks, AI-driven detection technologies, and public sentiment within these countries, the study highlights the uneven preparedness of BRICS nations in addressing the complexities posed by digital misinformation. While some, notably China and Russia, have made significant strides in technological innovation and regulatory enforcement, others, such as Brazil and South Africa, exhibit nascent efforts that remain mired in political and infrastructural challenges. Furthermore, the research underscores the imperative for organizations to adapt their strategic, operational, and crisis management frameworks to the evolving digital ecosystem, integrating cutting-edge tools for deepfake detection and establishing robust internal controls. The study concludes that while Western nations have made notable progress in legislative and technological responses, the global landscape remains fragmented and lacks cohesive international standards. Consequently, the managerial imperative is clear: to ensure organizational resilience and ethical media practices, both state and non-state actors must collaborate across borders, creating a unified front against the pernicious spread of digital disinformation.

Keywords Deepfake technology · Digital manipulation · BRICS nations · Artificial intelligence · Legal frameworks · Misinformation · Technological innovation · Public awareness

K. Dahiya (✉)
Manipal University Jaipur, Jaipur, Rajasthan, India
e-mail: kavitadahiya63@gmail.com

U. Sharma
Chandigarh Group of Colleges Jhanjeri, Mohali, India
e-mail: umang.hr@cgc.ac.in

S. Taneja et al. (eds.), *Navigating the Deepfake Conundrum: A Manager's Roadmap*, Information Systems Engineering and Management 44,
https://doi.org/10.1007/978-3-031-90742-5_16

1 Introduction

In the context of an ever-evolving global landscape characterized by intricate interdependencies and multifaceted challenges, the BRICS nations—Brazil, Russia, India, China, and South Africa—represent a consortium of emerging powers whose influence is both profound and multifarious [70]. These countries, each distinguished by its unique historical trajectory, socio-political milieu, and economic paradigm, have transcended their regional confines to assert a collective voice on the global stage, articulating aspirations for a more equitable and multipolar world order [42, 69, 82]. The sheer magnitude of their demographic, economic, and geopolitical significance imbues them with an unparalleled capacity to shape the contours of global governance, trade, and technological innovation [92].

Inextricably linked by their shared commitment to fostering an alternative narrative to the hegemonic structures of traditional Western dominance, the BRICS bloc is a testament to the evolving dynamics of global power [14, 32, 68]. Yet, while they collectively represent a formidable force, their divergent paths—marked by a confluence of developmental challenges, socio-economic disparities, political volatility, and environmental exigencies—underscore the complex and often discordant nature of their alliance. The existential paradox of their cooperative engagement lies in their simultaneous pursuit of both national self-interest and regional/global solidarity, often navigating the precarious terrain between convergence and divergence [39, 40, 41, 91].

Each nation within the BRICS grouping faces a unique constellation of obstacles, yet all are bound by a common pursuit of sustainable development, technological advancement, and enhanced geopolitical agency. Brazil, with its expansive natural resources and strategic emphasis on environmental stewardship, confronts the paradox of economic growth juxtaposed with ecological preservation [35, 63]. Russia, a resurgent geopolitical actor, asserts itself through its strategic dominance in energy, military might, and regional influence, even as it grapples with the repercussions of economic isolation and political centralization. India, an emerging technological juggernaut, balances its burgeoning IT and service sectors with the challenges of deep-seated socio-economic inequality and infrastructural bottlenecks that impede its full potential. China, the behemoth of the group, emerges as both a beacon of manufacturing prowess and an epicenter of technological innovation, even as it contends with its aging demographic and the ramifications of an increasingly fraught international relations landscape. South Africa, the gateway to the African continent, continues to assert its role as a regional leader while navigating the intricacies of post-apartheid economic restructuring, unemployment, and socio-political discord [48, 49, 87].

In this confluence of national imperatives and collective aspirations, the BRICS nations' efforts to recalibrate the global order—both economically and politically—present a nuanced tapestry of opportunities and challenges. The very nature of their alliance underscores the tension between the desire for global integration and the protection of national sovereignty, between the pursuit of innovation and the acknowledgment of historical inequities. As such, the trajectory of BRICS' influence will invariably hinge upon their ability to harmonize these divergent elements, simultaneously embracing the opportunities of a rapidly globalizing world while navigating the complexities of an interconnected, yet deeply fragmented, international system.

This emergent bloc stands not only as a counterbalance to traditional power structures but also as a harbinger of an evolving geopolitical and economic paradigm, one that holds the potential to redefine global power dynamics in the coming decades. Their collective role in shaping the future of global trade, security, technological advancements, and environmental governance will, undoubtedly, be pivotal in determining the contours of the twenty-first century's international order.

In the nascent digital age, where the convergence of technology and human cognition is increasingly becoming indistinguishable, the advent of synthetic media—most notably in the form of deepfake technology—presents unprecedented challenges and ethical conundrums for organizational leadership. Deepfakes, a sophisticated byproduct of advancements in artificial intelligence (AI) and machine learning, allow the generation of hyper-realistic yet entirely fabricated audiovisual content. These digital manipulations have shattered the traditional paradigms of truth, trust, and authenticity, which have long been cornerstones of effective leadership [84]. As business leaders navigate an increasingly volatile and complex environment, the ability to discern genuine content from its falsified counterpart has become integral to safeguarding corporate reputation, ensuring transparent communication, and maintaining public trust [34, 76, 85, 86].

The implications of deepfake technology on leadership are manifold. At the heart of the issue lies the erosion of the very notion of veracity, an essential tenet of leadership in any context—be it corporate, political, or social. Historically, leaders have relied on transparent and verifiable communication as the bedrock of their legitimacy. Whether in the boardroom, in public speeches, or through digital channels, leaders have expected that their words and actions would be perceived as authentic representations of their intent and character [45, 58]. However, in an era defined by deepfake technology, the very fabric of trust and credibility is under siege. The rise of manipulated video and audio content introduces a layer of ambiguity that makes the task of leadership significantly more complex, as the veracity of even the most seemingly innocuous piece of information can no longer be taken for granted [6, 24].

The ethical dilemmas precipitated by deepfake technology are equally profound. In a leadership context, the use of such technology is fraught with peril—both for the leader themselves and for the broader organizational ecosystem. Leaders, who are typically viewed as paragons of integrity and decision-making acumen, may find themselves enmeshed in ethical quandaries related to the potential misuse of deepfake technology. The temptation to employ such technology for personal or organizational gain—be it for political influence, market manipulation, or reputational enhancement—presents not only a risk to the ethical standards of the individual leader but also to the credibility of the institutions they represent. Furthermore, the malicious deployment of deepfakes, such as the creation of fabricated videos that purport to show leaders making controversial statements or engaging in unethical behavior, has the potential to destabilize an organization, fracture employee morale, and ignite public scandals.

From a managerial perspective, deepfake technology introduces an array of strategic challenges. The proliferation of digital misinformation necessitates a paradigm shift in how leaders approach risk management, communication strategies, and crisis mitigation. Traditional forms of leadership, which often relied on trust-building and transparent communication, are now tested by the pervasive nature of digital manipulation. In the corporate world, where brand equity and consumer confidence are fragile, the consequences of a deepfake-related scandal can be disastrous. Consequently, leadership must evolve to encompass a multifaceted approach to safeguarding an organization's integrity—spanning crisis communication, brand management, and the cultivation of a digital-savvy workforce. The increasing sophistication of deepfake technology means that the role of leadership must now extend beyond traditional managerial functions to include the implementation of robust digital governance frameworks capable of mitigating the risks associated with synthetic media.

One of the most striking aspects of deepfake technology's challenge to leadership is its democratization. In the past, the ability to manipulate media was confined to a select few with access to advanced technological tools and resources. However, the rapid advancement of AI, coupled with the availability of sophisticated deepfake creation tools, has democratized the ability to produce highly convincing forged content [77, 97]. This democratization significantly amplifies the scope of leadership challenges, as leaders are now confronted not only with threats from organized adversaries but also from individuals or small groups with the requisite technical skills. Moreover, the relative ease with which deepfakes can be disseminated through social media platforms and other online channels exacerbates the speed at which misinformation can spread, further complicating the task of leadership [15, 81].

As the use of deepfake technology proliferates, the ethical responsibility of leaders becomes ever more pressing. On one hand, leaders must grapple with the potential for deepfakes to be weaponized for malicious purposes, such as cyberattacks, corporate espionage, or political destabilization. On the other hand, they must contend with the temptation of leveraging this technology to further organizational agendas. The ethical labyrinth is further complicated by the fact that deepfakes, by their very nature, are designed to appear indistinguishable from reality. This creates an environment where manipulation can occur with little to no immediate recognition, making it difficult for leaders to ascertain the truthfulness of information presented to them. This not only complicates decision-making processes but also necessitates a rethinking of the ethical frameworks that guide leadership in the digital age [2, 26].

In this context, the challenge for leadership is not simply one of responding to the threat of deepfakes but also of anticipating and proactively addressing the implications of their use. Organizational leaders must cultivate a heightened sense of digital literacy and media discernment, equipping themselves with the tools and knowledge necessary to identify, challenge, and mitigate the potential risks posed by synthetic media [38, 47]. They must also play an instrumental role in fostering a corporate culture that prioritizes ethical decision-making, transparency, and accountability—values that stand in direct opposition to the manipulative potential of deepfake technology. Moreover, leadership must embrace a forward-thinking approach to crisis management, ensuring that robust mechanisms are in place to rapidly detect and address any instances of deepfake manipulation that may arise within their organization or the broader public domain [22, 52].

In addition to these challenges, there is a broader question of regulation and accountability in the realm of deepfakes. While technological advancements often outpace the legislative frameworks designed to govern them, the ethical and societal consequences of deepfake manipulation require urgent attention from policymakers and regulatory bodies. The governance of deepfake technology is a topic that transcends the purview of any single organization, industry, or nation. It is a global issue that necessitates a coordinated, multifaceted response to ensure that the use of such technology is held to ethical standards that prevent its malicious exploitation [51, 71]. In this regard, leaders not only need to focus on internal organizational responses but also engage in broader dialogues surrounding the ethical governance of AI and synthetic media.

Ultimately, the leadership challenges posed by deepfake technology are emblematic of the broader ethical dilemmas that define the 21st-century business landscape. As organizations become increasingly digitized and interconnected, leaders are tasked with navigating a world in which technological innovation constantly shifts the boundaries of what is possible—and what is ethical [9, 29]. Deepfake technology, with its potential to disrupt truth, transparency, and accountability, forces leaders to reconsider the very foundations of their roles. The ability to lead effectively in the age of deepfakes will require not only an understanding of the technological underpinnings of AI and machine learning but also a commitment to the ethical stewardship of the organizations and individuals they serve.

In conclusion, the intersection of leadership and deepfake technology is one fraught with complexity, uncertainty, and risk. While the potential for innovation and efficiency is clear, the ethical dilemmas are inescapable. For leaders in this new age, the challenge is twofold: to harness the benefits of technological advancements while simultaneously safeguarding against their misuse. In doing so, they must grapple with difficult questions regarding integrity, accountability, and trust in a world where authenticity is increasingly elusive. The future of leadership, therefore, will be defined not only by a capacity for technological adaptability but also by a steadfast commitment to the ethical principles that underpin effective and responsible governance.

2 Literature Review

In recent years, the intersection of deepfake technology and leadership has become an emergent subject of critical discourse. The evolving landscape of artificial intelligence has fundamentally altered the dynamics of communication, eroding the previously sacrosanct boundaries between truth and falsity [79, 89]. Deepfakes, as a manifestation of machine learning algorithms, enable the creation of hyper-realistic yet entirely fabricated audiovisual content, which is often indistinguishable from genuine media. This technological advancement presents a multifaceted dilemma for leaders, forcing a reconsideration of how authority, credibility, and authenticity are constructed and maintained in an era of rampant digital manipulation [12, 59].

The implications of deepfake technology extend beyond mere technical innovation; they resonate deeply within the ethical framework that governs leadership practice. As leaders are traditionally entrusted with the stewardship of truth, transparency, and accountability, the proliferation of synthetic media forces a re-evaluation of how these principles are upheld in a landscape where the boundary between fact and fiction has become increasingly nebulous. The integrity of leaders, once anchored in their ability to communicate reliably and authentically, is now subject to the corrosive effects of fabricated media, thus challenging conventional paradigms of trust and perception. As a result, leadership becomes inextricably linked to the ability to discern and authenticate the veracity of information, particularly in an age where manipulated content can be disseminated with alarming ease [4, 90].

At the core of this ethical quandary is the issue of manipulation. Deepfake technology, when wielded with malicious intent, becomes a potent tool for deception, capable of distorting public perception, undermining reputations, and influencing the outcomes of critical decisions. The implications for organizational leadership are profound, as leaders must now navigate a realm where their words, actions, and even their very image can be artificially reconstructed and weaponized against them. The concept of "leadership authenticity" has thus evolved into a precarious balancing act—leaders are compelled to defend not only their ideas and decisions but also their very identities, which may be vulnerable to misrepresentation in the public sphere [23, 60].

Moreover, the ethical dilemmas that emerge from the integration of deepfake technology into leadership practices are not confined to external threats alone. Internally, organizations must grapple with the implications of employing such technologies in communication strategies. The question arises whether the use of synthetic media could ever be justified as a tool for leadership communication, or if its mere existence poses an insurmountable ethical challenge. In a landscape where perception is paramount, leaders must also contend with the potential for internal disillusionment, as the line between genuine leadership and artificially constructed personas becomes increasingly difficult to delineate [3, 19].

In this context, the ethical risks of deepfakes are not limited to their potential to deceive the public or undermine trust. They also raise existential questions about the nature of authority in the digital age. Traditional leadership models, which rely on consistency, transparency, and authenticity, may find themselves increasingly at odds with the digital realities of the twenty-first century. As synthetic media becomes more sophisticated, leaders must develop new frameworks for ethical decision-making, ensuring that their communication remains grounded in truth while navigating a media landscape rife with disinformation [55].

Additionally, the challenge of managing deepfake technology extends to the realm of crisis management. The ability to respond effectively to the malicious use of deepfakes is an essential skill for modern leaders. Whether confronting a crisis instigated by the deliberate spread of a fabricated video or defending the organization against reputational harm caused by digital manipulation, leaders must develop strategies for identifying, authenticating, and countering synthetic media. The rapid dissemination of digital content, combined with the difficulty of detecting deepfakes, places leaders in a perpetual state of vigilance, where the authenticity of any piece of information is perpetually questioned [25].

In parallel, the psychological effects on both leaders and followers cannot be overlooked. The constant threat of digital manipulation can create an atmosphere of mistrust, eroding the psychological contract between leaders and their followers [16]. Leaders, once revered for their ability to convey the vision and inspire confidence, must now contend with a public that is increasingly sceptical of the authenticity of any media they encounter. For followers, the proliferation of synthetic media fosters a sense of disillusionment, as they may begin to question the very nature of leadership itself, doubting whether any representation of a leader—whether visual, auditory, or textual—can indeed be trusted [10].

In this evolving paradigm, the very concept of leadership is transforming. The traditional view of leaders as steadfast figures of authenticity and trustworthiness is being challenged by the disruptive potential of deepfake technology [33, 43]. Leaders are no longer merely tasked with managing resources or making decisions; they must now navigate a complex web of digital media, where their every word and image can be distorted, manipulated, and weaponised. In response, they must develop new strategies to safeguard their credibility, maintain organisational cohesion, and protect the ethical integrity of their actions.

The societal implications of deepfakes also demand careful consideration. In a broader sense, the integration of synthetic media into leadership practices speaks to a profound shift in the way power and influence are exercised in the modern world. Leadership, once bound by the limitations of physical presence and face-to-face interaction, is now extended into a digital realm where reality is often mutable. This raises important questions about the future of democracy, governance, and public discourse, as deepfake technology has the potential to destabilize established norms and erode trust in public institutions [57].

In conclusion, the advent of deepfake technology presents a profound challenge to traditional leadership paradigms. As leaders in both the private and public sectors grapple with the ethical, reputational, and operational implications of this disruptive technology, they must develop new approaches to maintaining authenticity, transparency, and accountability. Deepfake technology forces a rethinking of the foundational principles of leadership, calling for an evolution of ethical frameworks and communication strategies that can withstand the growing threat of digital manipulation. As organizations and leaders continue to adapt to this new reality, it will be imperative to balance technological innovation with an unwavering commitment to integrity and truthfulness.

3 Research Methodology

The research methodology underpinning the composition of this chapter is rooted in a multidimensional and interdisciplinary approach, combining qualitative analyses, comparative frameworks, and a nuanced interpretation of geopolitical, technological, and regulatory developments related to deepfake technology. This methodology draws on a range of academic disciplines, including political science, digital ethics, international relations, information technology, and legal studies, creating a comprehensive and integrative examination of the subject matter. Central to this research was a thematic literature review, meticulously conducted across various fields such as digital media manipulation, artificial intelligence (AI), deepfake technologies, and international law. The objective of this review was to contextualize the deepfake phenomenon within the broader discourse on emerging technologies and their societal implications. Through critical analysis of existing academic sources, policy reports, and technological whitepapers, the research aimed to map the evolving global narrative surrounding deepfakes, particularly focusing on their geopolitical ramifications and the ethical dilemmas they introduce.

The methodology also employed a comparative analysis framework to juxtapose the responses of BRICS nations with the global stance on deepfake regulation, technological development, and public awareness. BRICS nations were selected deliberately due to their emerging influence in global governance and their diverse socio-political and technological environments. This comparative analysis focused on assessing the state of deepfake regulation, AI advancements in detection, and the general level of public awareness within each BRICS country, before placing these findings in contrast with the responses from the rest of the world, particularly from the European Union, United States, and other developed economies. This allowed for a robust and dynamic comparison, illustrating the disparity and convergence in global approaches to deepfake challenges.

Integral to the methodology was the synthesis of qualitative data, which played a pivotal role in understanding the multifaceted nature of the deepfake issue. Data was drawn from governmental reports, legal frameworks, and policy briefs to examine the legislative landscapes within BRICS nations. Additionally, the evolving discourse on the ethical considerations of deepfake technologies was explored through the lenses of political theory and digital ethics. The research delved into how digital manipulation impacts democratic processes, social trust, and national security in the BRICS context, with a particular focus on state surveillance and information control. This intricate analysis provided a more holistic understanding of the challenges posed by deepfakes in the global geopolitical environment.

4 Data Analysis

These events in Table 1 illustrate the growing intersection between deepfake technology and leadership across various sectors, from politics and business to education and media. The emergence of deepfake-related issues necessitates a revaluation of ethical frameworks, crisis management strategies, and the role of technological safeguards in maintaining leadership integrity.

This framework, presented in Table 2, outlines the various aspects of leadership that must be considered in the age of deepfakes, focusing on the practical and ethical challenges leaders face.

Summary of BRICS nation is global development in given in Table 3.

In the contemporary digital age, the advent of deepfake technology has precipitated an unprecedented array of challenges and ethical quandaries, compelling nations across the globe to grapple with the implications of artificial intelligence-driven media manipulation as shown in Table 4. As this technology proliferates, the responses of individual nations, particularly those within the BRICS consortium, exhibit a heterogeneous spectrum of preparedness, regulatory frameworks, and technological advancements. These countries, marked by disparate socio-political landscapes, economic structures, and technological infrastructure, find themselves at varying stages of engagement with the deepfake phenomenon. While the BRICS bloc, as a collective, has made significant strides in certain domains, their efforts remain nuanced and, in many instances, sporadic when juxtaposed with the broader global context. The table below provides a comparative analysis, highlighting the current state of deepfake legislation, technological development, and public awareness within the BRICS nations, alongside the strategies and responses from the rest of the world, thus offering a panoramic view of the international landscape in the face of this emergent threat.

5 Managerial Implications

The managerial implications of the rise of deepfake technology are profound, demanding a recalibration of both strategic and operational frameworks within organisations, especially for leaders operating in regions represented by BRICS nations. As digital manipulation escalates in sophistication, the capacity for disinformation, reputational damage, and erosion of stakeholder trust amplifies, placing an imperative on executives to enhance their digital literacy and crisis management protocols. In the context of BRICS nations, where regulatory frameworks and technological capabilities are in varying stages of development, managers must navigate an increasingly fragmented landscape of legislation, cybersecurity measures, and public sentiment. The absence of cohesive global standards further compounds the challenge, requiring firms to adopt a proactive stance in mitigating risks associated with deepfakes by

Table 1 Recent advancements

Source	Event/issue	Impact on leadership	Description
[99]	Political scandal in country X involving deepfake video of a leading politician	Erosion of trust in leadership	A deepfake video purporting to show a prominent politician making controversial statements goes viral, sparking debates on the role of AI in political manipulation and disinformation
[80]	Corporate crisis: CEO's deepfake video sparks social media backlash	Reputation management challenges	A CEO's likeness is used in a deepfake video that falsely implicates them in unethical business practices. The company must navigate crisis communication and rebuild stakeholder trust
[64]	Social media platforms introduce deepfake detection tools	Technological response to leadership integrity	Leading social media companies announce new AI-driven tools aimed at detecting deepfakes. Leaders within these firms face the challenge of ensuring these tools are effective and ethical
[93]	Educational institutions begin using deepfakes for simulation and training	Innovative leadership in education	Universities experiment with deepfake technology in leadership training, using AI-generated scenarios to teach decision-making in crisis situations. Ethical concerns over authenticity arise
[53]	CEO's public statement on combating misinformation using AI-driven technologies	Ethical leadership in technology use	The CEO of a global tech firm announces a new initiative aimed at combating deepfakes, pledging transparency in AI usage to preserve trust in the company and its leadership
[18]	Deepfake documentary on political leaders sparks ethical debate	Leadership accountability and media responsibility	A documentary series using deepfake technology to reconstruct speeches and actions of political figures prompts widespread debate on the responsibility of media and leadership ethics
[17]	Legal battle: deepfake videos used in corporate espionage case	Legal and ethical dilemmas for leadership	A multinational corporation sued a rival firm for using deepfake videos to steal intellectual property. This case sets legal precedents for how deepfakes are handled in corporate disputes

implementing robust internal controls, conducting regular cybersecurity audits, and engaging in continuous stakeholder education.

Moreover, the managerial responsibility extends beyond the corporate perimeter, necessitating an alignment with international best practices in media ethics and data integrity. In regions with underdeveloped legal frameworks, such as Brazil or South Africa, companies must act as stewards of ethical media use, fostering a culture of transparency and accountability that transcends the limitations of existing laws.

Table 2 Related theories

Theoretical aspect	Description	Key components/ concepts	Relevance to leadership	Source
Leadership ethics in the digital era	Focus on the ethical considerations that leaders must navigate in an era dominated by digital manipulation	Transparency, accountability, truth, trustworthiness, responsibility	Leaders need to maintain ethical standards in a world where technology can manipulate truth and influence perception	[56, 88]
Media literacy and critical thinking	Emphasises the importance of media literacy and the ability to assess digital content critically	Media literacy, misinformation, deepfakes, discernment	Leaders must possess a high degree of media literacy to assess the veracity of information and protect their organizations from harm	[74]
Trust and reputation management	Understanding how trust and reputation are affected by digital misinformation, especially deepfakes	Trust, reputation, credibility, public perception	Leaders must act to maintain their personal and organizational reputation in the face of manipulated media	[73, 78]
Decision-making under uncertainty	How leaders make informed decisions despite the growing uncertainty due to digital misinformation	Risk assessment, decision-making, uncertainty, resilience	Effective decision-making requires managing the unknowns of misinformation, bias, and manipulated media	[37, 95]
Ethical technology use and responsibility	The role of leaders in guiding the ethical use of new technologies, including the regulation of deepfakes	Regulation, digital ethics, policy-making, governance	Leaders are responsible for implementing policies that address the risks posed by deepfakes and other advanced technologies	[31, 94]
Organizational leadership and influence	The ability of leaders to influence organizational culture and guide responses to ethical challenges	Influence, leadership styles, organizational culture, ethical climate	Leadership shapes how an organization responds to challenges posed by deepfakes and manipulative technologies	[30, 96]

(continued)

Table 2 (continued)

Theoretical aspect	Description	Key components/ concepts	Relevance to leadership	Source
Public perception and social responsibility	How leaders manage the public's perception of their brand or organization in light of deepfake challenges	Public relations, social responsibility, brand image, corporate social responsibility (CSR)	Leaders must ensure their public statements and actions reflect social responsibility and truthfulness, counteracting deepfake misuse	[62, 66]
Crisis management and communication	Handling crises triggered by digital misinformation, including deepfakes, through effective communication	Crisis communication, media management, transparency, rapid response	Leaders must be prepared to communicate effectively in crises to maintain credibility and control over the narrative	[5, 98]
Technological literacy for leaders	The importance of leaders having an understanding of digital technologies like deepfakes	Digital literacy, technological awareness, AI, machine learning, deepfake creation	Leaders must understand the implications of technology to ensure they can make informed decisions and protect against manipulation	[3, 27]

Conversely, in technologically advanced regions like China and Russia, where AI detection technologies are being rapidly deployed, organizations must prioritize investment in cutting-edge tools to detect and neutralize deepfakes while ensuring compliance with nascent but evolving regulatory landscapes.

In terms of operational strategy, the pervasive threat of deepfakes calls for an integration of AI-driven solutions within the corporate infrastructure, particularly in media monitoring, brand protection, and crisis communication channels. From a broader managerial perspective, there is a compelling need to foster cross-industry collaboration to address the multifaceted nature of deepfake challenges. This entails not only leveraging technological expertise but also influencing public policy to create a more coherent and equitable global approach to digital manipulation. The overarching managerial imperative is to ensure organizational resilience in an era where the digital landscape is increasingly marred by the distortion of reality, necessitating a forward-thinking, agile, and ethically grounded approach to leadership in the face of such existential risks.

Table 3 BRICS nations: shaping global trends and opportunities in the 21st century

BRICS nation	Position in global developments	Key contributions/ areas of influence	Challenges	Opportunities	Source
Brazil	Emerging global leader in sustainable development and agribusiness	Agriculture, renewable energy, climate change policies	Political instability, economic inequality, infrastructure deficits	Expanding renewable energy sector, sustainable agriculture, environmental policies	[54, 74]
Russia	Major global player in geopolitical issues and energy markets	Energy (oil and gas), defense, security, space exploration	Sanctions, economic diversification challenges, international isolation	Energy exports, military technology, global influence in Eastern Europe & Central Asia	[47]
India	Rising economic and technological hub, leading in IT and services	Information technology, space exploration, pharmaceuticals, defense	Social inequality, infrastructure bottlenecks, political fragmentation	IT industry growth, renewable energy, manufacturing sector, international diplomacy	[36, 61]
China	World's second-largest economy, leading in manufacturing and technology	Manufacturing, artificial intelligence, infrastructure (belt and road initiative)	Aging population, environmental issues, international relations tensions	Global trade leadership, technological innovation, green energy transition	[11, 20]
South Africa	Key African economy, focusing on trade, mining, and regional leadership	Mining (precious metals), tourism, regional peacekeeping, Infrastructure development	Unemployment, inequality, political unrest	Regional leadership, tourism, mining innovation, renewable energy projects	[1, 67]

6 Conclusion

In conclusion, the emergence of deepfake technology as a formidable force within the digital ecosystem represents a profound paradigm shift, challenging traditional notions of truth, trust, and media integrity. This transformative phenomenon has precipitated a convergence of technological, ethical, and geopolitical dilemmas, compelling both state and non-state actors to confront the complexities inherent in regulating and mitigating its consequences. The BRICS nations, with their disparate developmental trajectories, socio-political dynamics, and technological capacities,

Table 4 Global dynamics of deepfake technology: a comparative analysis of BRICS Nations and the world

Aspect	Brazil	Russia	India	China	South Africa	Rest of the world
Legislation and regulation [46, 83]	Lacks comprehensive legislation on deepfake regulation. Some initiatives are underway, but there is no robust framework	Russia has laws against the creation and distribution of deepfakes, but enforcement remains limited	India has a patchy approach with emerging frameworks, though still lagging in terms of enforcement	China is actively developing policies to curb deepfakes, but enforcement varies across regions	South Africa lacks specific legislation on deepfakes but is beginning to explore its implications	The EU, the US, and other developed nations are increasingly enacting laws and guidelines to regulate deepfakes, but enforcement is still inconsistent
Technological development [8, 72]	Brazil's technological response to deepfakes is minimal, with limited investment in AI/ML for detection	Russia has significant cyber capabilities and is working on tools for detecting deepfakes, but faces challenges due to resource limitations	India is investing heavily in AI, but deepfake detection technologies are still in nascent stages	China leads in AI and is developing sophisticated deepfake detection technologies, but state surveillance poses ethical concerns	South Africa is slowly catching up in AI and cybersecurity but lags behind other BRICS nations	The US and EU have advanced research in AI for deepfake detection, with many tech companies developing solutions. However, the fast-evolving technology poses an ongoing challenge
Public awareness [7, 44]	Public awareness of deepfakes is still low, though growing in certain circles	Russia has a moderate level of awareness, especially among the political elite, but general public awareness remains limited	Awareness is increasing, but the general public is still not fully cognizant of the threat posed by deepfakes	China has relatively high awareness due to state-driven campaigns, but also state-driven narrative control	Awareness in South Africa is low, though there is growing concern about the impact on elections	Public awareness is highest in Western countries, with many campaigns dedicated to educating the public on the dangers of deepfakes

(continued)

Table 4 (continued)

Aspect	Brazil	Russia	India	China	South Africa	Rest of the world
Impact on society and politics [50, 75]	Growing concern over deepfakes in politics, especially during elections	Deepfakes are weaponized for political purposes, with significant disinformation campaigns, particularly in the context of international relations	Deepfakes have started to impact political discourse, especially in elections and media	Deepfakes are used in political propaganda and information warfare, particularly regarding Xinjiang and Hong Kong issues	Deepfakes are emerging as a threat to elections and political stability	In Western democracies, deepfakes are increasingly used for political manipulation, and awareness campaigns focus on their societal impacts
Collaboration and international response [28, 65]	Brazil is participating in international discussions but lacks a clear leadership role in anti-deepfake efforts	Russia is a participant in international cybersecurity dialogues but often operates in opposition to Western regulations	India is part of global efforts but remains passive in formulating robust international policies on deepfakes	China is actively involved in international tech forums but often pushes for state-centric regulation models that conflict with Western ideals	South Africa is starting to collaborate more with global organizations but is not a key player in international deepfake regulation	Developed countries like the US, EU, and Japan have robust international collaboration in regulating deepfakes, with some pushing for a global treaty
Technological ethics and research [13, 21]	Brazil is beginning to engage in discussions about ethical AI and deepfakes, but research is limited	Ethical concerns are being raised, but state control and political interests often overshadow research on deepfakes	India is fostering tech research but struggles with an ethical framework for AI and deepfake technology	China emphasizes state control and often ties technological advancements with surveillance, raising ethical concerns	South Africa is at an early stage of ethical discussions but lacks funding and infrastructure for robust research	In the US and EU, ethical discussions around AI and deepfakes are prominent, with significant funding directed toward ethical AI research

exemplify a microcosm of the global challenge posed by deepfakes. While certain nations within this bloc, such as China and Russia, exhibit considerable technological prowess and governmental oversight in combating digital manipulation, others, like Brazil and South Africa, face significant institutional and infrastructural barriers in establishing coherent legal frameworks and detection mechanisms.

The comparative analysis highlights not only the disparities in the readiness and response of BRICS nations to deepfake threats but also underscores the broader global inequities in addressing the risks of digital misinformation. The rest of the world, particularly in Western contexts, has made significant strides in developing legal standards, regulatory oversight, and technological innovations to combat deepfake proliferation. However, the effectiveness of these measures remains contingent upon international cooperation, standardization, and the capacity to adapt to the ever-evolving nature of artificial intelligence and its applications in media manipulation.

For leaders and organizations, the implications are multifaceted and urgent. As deepfake technology continues to evolve with alarming velocity, so too must the strategic frameworks and operational protocols designed to safeguard public trust, brand integrity, and organizational reputation. The managerial responsibility is not confined to the implementation of reactive measures but extends to proactive engagement with emerging technological tools, legal reforms, and ethical guidelines to ensure resilience in the face of this existential threat. Ultimately, the global community must foster a collective resolve to address the deepfake dilemma, not as isolated entities but as a united front, navigating the confluence of technological innovation, regulatory oversight, and societal ethics in a manner that upholds the sanctity of truth and the integrity of digital media.

References

1. Acar, E., Sarıyer, G., Jain, V., Ramtiyal, B.: Discovering hidden associations among environmental disclosure themes using data mining approaches. Sustainability (Switzerland) **15**(14) (2023). https://doi.org/10.3390/su151411406
2. Adelodun, B., Agbelusi, O.O., Soma, T., Odey, G., Adeyi, Q., Kumar, P., Ajibade, F.O., Goala, M., Silva, L.F.O., Mostafa, Y.S., Singh, R., Choi, K.S., Eid, E.M.: Rethinking food loss and waste to promote sustainable resource use and climate change mitigation in agri-food systems: a review. Waste Manage. Res. (2024). https://doi.org/10.1177/0734242X241257655
3. Ahmad, I., Sharma, S., Kumar, R., Dhyani, S., Dumka, A.: Data analytics of online education during pandemic health crisis: a case study. In: 2nd Edition of IEEE Delhi Section Owned Conference, DELCON 2023—Proceedings (2023). https://doi.org/10.1109/DELCON57910.2023.10127423
4. Akberdina, V., Kumar, V., Kyriakopoulos, G.L., Kuzmin, E.: Editorial: what does industry's digital transition hold in the uncertainty context? In: Lecture Notes in Information Systems and Organisation, vol. 61 LNISO, pp. 1–4. Springer Science and Business Media Deutschland GmbH (2023). https://doi.org/10.1007/978-3-031-30351-7_1
5. AL-Huqail, A.A., Singh, R., Širić, I., Kumar, P., Abou Fayssal, S., Kumar, V., Bachheti, R.K., Andabaka, Ž., Goala, M., Eid, E.M.: Occurrence and health risk assessment of heavy metals in Lychee (Litchi chinensis Sonn., Sapindaceae) fruit samples. Horticulturae **9**(9) (2023). https://doi.org/10.3390/horticulturae9090989

6. Anand, A., Singh, M.P., Sah, S., Verma, A., Khatib, M.N., Pant, M., Zahiruddin, Q.S., Rustagi, S.: H5N1 avian influenza in USA: a call for vigilance in one health surveillance. Clin. Infect. Pract. **23**,(2024). https://doi.org/10.1016/j.clinpr.2024.100369
7. Ashish, P.R., Vashistha, R., Singh, R., Thakur, K.: Petroleum industry and its effect on climate. In: AIP Conference Proceedings, vol. 2978, no. 1. American Institute of Physics Inc. (2024). https://doi.org/10.1063/5.0183164
8. Asthana, N., Pal, K., Khan, A.A., Malik, A.: Novel biopolymeric materials potential utilization for environmental practices. J. Mol. Struct. **1311**,(2024). https://doi.org/10.1016/j.molstruc.2024.138390
9. Ayad, H., Shuaib, M., Hossain, M.E., Haseeb, M., Kamal, M., ur Rehman, M.: Re-examining the environmental kuznets curve in MENA countries: is there any difference using ecological footprint and CO2 emissions? Environ. Model. Assess. https://doi.org/10.1007/s10666-024-09977-7
10. Badhotiya, G.K., Avikal, S., Soni, G., Sengar, N.: Analyzing barriers for the adoption of circular economy in the manufacturing sector. Int. J. Product. Perform. Manag. **71**(3), 912–931 (2022). https://doi.org/10.1108/IJPPM-01-2021-0021
11. Balaji, V., Kaliappan, S., Madhuvanesan, D.M., Ezhumalai, D.S., Boopathi, S., Patil Pravin, P., Mani, S.: Combustion analysis of biodiesel-powered propeller engine for least environmental concerns in aviation industry. Aircr. Eng. Aerosp. Technol. **94**(5), 760–769 (2022). https://doi.org/10.1108/AEAT-11-2021-0344
12. Bansal, N., Taneja, S., Ozen, E.: Green financing as a bridge between green banking strategies and environmental performance in Punjab, India. Int. J. Sustain. Dev. Plan. **18**(10), 3155–3167 (2023). https://doi.org/10.18280/IJSDP.181017
13. Bhandari, G., Dhasmana, A., Chaudhary, P., Gupta, S., Gangola, S., Gupta, A., Rustagi, S., Shende, S.S., Rajput, V.D., Minkina, T., Malik, S., Slama, P.: A perspective review on green nanotechnology in agro-ecosystems: opportunities for sustainable agricultural practices & environmental remediation. Agriculture (Switzerland) **13**(3) (2023). https://doi.org/10.3390/agriculture13030668
14. Bhardwaj, M., Prakash, N., Malhotra, R.K.: Energy sector and carbon trading in the future: a reference to India. In: 2024 International Conference on Smart Devices, ICSD 2024 (2024). https://doi.org/10.1109/ICSD60021.2024.10751388
15. Bhat, R., Gupta, M., Choudhary, T.F., Singh, A.K.: Modelling the effects of climatic variables and political instability on the agricultural sector in South Asian countries. In: Building Climate Neutral Economies Through Digital Business and Green Skills. IGI Global (2024). https://doi.org/10.4018/979-8-3693-5673-9.ch001
16. Bhatnagar, M., Sehajpal, S.: Leveraging AI for financial sustainability in healthcare. In: Ercan, O., Azad, S., Sanjay, T., Rajendra, R., Paulo, D.J. (eds.) Sustainability, Innovation, and Consumer Preference, pp. 411–426. IGI Global Scientific Publishing (2024). https://doi.org/10.4018/979-8-3693-9699-5.ch017
17. Bhatti, K.K., Negi, A.: Green store attributes and infrastructure: A connecting link between retailers and consumers to bring sustainability. Int. J. Civ. Eng. Technol. **9**(11), 190–198 (2018)
18. Bohra, N.S., Bansal, S.S., Johri, A., Kathari, S.: Evaluating infrastructure fund performance in India: a study of thematic investing. In: Issues of Sustainability in AI and New-Age Thematic Investing. IGI Global (2024). https://doi.org/10.4018/979-8-3693-3282-5.ch002
19. Davuluri, S.K., Alvi, S.A.M., Aeri, M., Agarwal, A., Serajuddin, M., Hasan, Z.: A security model for perceptive 5G-powered BC IoT associated deep learning. In: 6th International Conference on Inventive Computation Technologies, ICICT 2023—Proceedings, pp. 118–125. https://doi.org/10.1109/ICICT57646.2023.10134487
20. Deng, Q., Usman, M., Irfan, M., Haseeb, M.: The role of financial inclusion and tourism in tackling environmental challenges of industrialization and energy consumption: redesigning Sustainable Development Goals policies. Nat. Res. Forum (2024). https://doi.org/10.1111/1477-8947.12522
21. Devade, K., Singh, P.K., Kumar, S., Kumar, H., Prasad, B., Rao, A.L.N., Sankhyan, A.: Green nanotechnology based sustainable energy solutions and environmental impacts. In: E3S Web of Conferences, vol. 511. EDP Sciences (2024). https://doi.org/10.1051/e3sconf/202451101031

22. Dhayal, K.S., Agrawal, S., Agrawal, R., Kumar, A., Giri, A.K.: Green energy innovation initiatives for environmental sustainability: current state and future research directions. Environ. Sci. Pollut. Res. **31**(22), 31752–31770 (2024). https://doi.org/10.1007/s11356-024-33286-x
23. Diwakar, M., Shankar, A., Chakraborty, C., Singh, P., Arunkumar, G.: Multi-modal medical image fusion in NSST domain for internet of medical things. Multimed. Tools Appl. **81**(26), 37477–37497 (2022). https://doi.org/10.1007/s11042-022-13507-6
24. Goyal, S., Kaur, A., Sharma, V., Batra, N.: IMHS: an IoT-based machine learning healthcare system. In: 2024 2nd International Conference on Disruptive Technologies, ICDT 2024, pp. 916–919 (2024). https://doi.org/10.1109/ICDT61202.2024.10489127
25. Gusain, I., Sharma, S., Debarma, S., Kumar Sharma, A., Mishra, N., Prakashrao Dahale, P.: Study of concrete mix by adding Dolomite in conventional concrete as partial replacement with cement. Mater. Today: Proceed. **73**, 163–166 (2023). https://doi.org/10.1016/j.matpr.2022.09.583
26. Hossain, M.R., Dash, D.P., Das, N., Hossain, M.E., Haseeb, M., Cifuentes-Faura, J.: Do trade-adjusted emissions perform better in capturing environmental mishandling among the most complex economies of the world? Environ. Model. Assess. (2024). https://doi.org/10.1007/s10666-024-09994-6
27. Husen, A., Bachheti, R.K., Bachheti, A.: Non-timber forest products: food, healthcare and industrial applications. In: Non-Timber Forest Products: Food, Healthcare and Industrial Applications. Springer International Publishing (2021). https://doi.org/10.1007/978-3-030-73077-2
28. Joshi, H., Nokshi, Chauhan, R., Rawat, R., Gupta, R.: Climate change: the impact and exploring the path forward. In: Proceedings of the 2nd International Conference on Intelligent and Innovative Technologies in Computing, Electrical and Electronics, ICIITCEE 2024. Institute of Electrical and Electronics Engineers Inc. (2024). https://doi.org/10.1109/IITCEE59897.2024.10467588
29. Joshi, M.N., Dixit, A.K., Saxena, S., Memoria, M., Choudhury, T., Sar, A.: A study of the application of AI & ML to climate variation, with particular attention to legal & ethical concerns. EAI Endorsed Trans. Internet of Things **10** (2024). https://doi.org/10.4108/eetiot.5468
30. Josphineleela, R., Jyothi, M., Natrayan, L., Kaviarasu, A., Sharma, M.: Development of IoT based health monitoring system for disables using microcontroller. In: Proceedings—7th International Conference on Computing Methodologies and Communication, ICCMC 2023, pp. 1380–1384 (2023). https://doi.org/10.1109/ICCMC56507.2023.10084026
31. Kadayat, Y., Sharma, S., Agarwal, P., Mohan, S.: Internet-of-things enabled smart health monitoring system using auto AI: a graphical tool of IBM Watson studio. Internet of Things, Part **F2482**, 427–445 (2024). https://doi.org/10.1007/978-981-97-0052-3_21
32. Khanna, R., Sharma, S., Czeh, T., Adhana, K.: Challenges, interventions, and strategies for higher education transformation: vision 2047. Lect. Notes Netw. Syst. 1075 LNNS, 637–655 (2025). https://doi.org/10.1007/978-981-97-6106-7_38
33. Kollipara, V.N.H., Kalakota, S.K., Chamarthi, S., Ramani, S., Malik, P., Karuppiah, M.: Timestamp based OTP and enhanced RSA key exchange scheme with SIT encryption to secure IoT devices. J. Cyber Secur. Mob. **12**(1), 77–102 (2023). https://doi.org/10.13052/jcsm2245-1439.1214
34. Kour, M., Taneja, S., Özen, E., Sood, K., Grima, S.: Financial Landscape Transformation Technological Disruptions (2025)
35. Kukreti, M., Sehajpal, S., Tiwari, R., Sood, K.: Driving Global Health and Sustainable Development Goals With Smart Technology. In: Mohit, K., Sabina, S., Rajesh, T., Kiran, S. (eds.), 1st ed. IGI Global Scientific Publishing (2025). https://doi.org/10.4018/979-8-3373-0240-9
36. Kumar, K., Chaudhary, S., Anandaram, H., Kumar, R., Gupta, A., Joshi, K.: Industry 4.0 and health care system with special reference to mental health. In: 2023 1st International Conference on Intelligent Computing and Research Trends, ICRT 2023 (2023). https://doi.org/10.1109/ICRT57042.2023.10146640
37. Kumar, M.A., Prasad, M.S.G., More, P., Christa, S.: Artificial intelligence-based personal health monitoring devices. In: Mobile Health: Advances in Research and Applications—Volume II. Nova Science Publishers, Inc. (2022)

38. Kumar, R., Goel, R., Singh, T., Mohanty, S.M., Gupta, D., Alkhayyat, A., Khanna, R.: Sustainable finance factors in indian economy: analysis on policy of climate change and energy sector. Fluct. Noise Lett. **23**(2) (2024). https://doi.org/10.1142/S0219477524400042
39. Kumar, R., Malhotra, R.K., Grover, N.: Data mining in credit scoring and future application. In: International Conference on Innovative Data Communication Technologies and Application, ICIDCA 2023—Proceedings, pp. 1096–1100 (2023). https://doi.org/10.1109/ICIDCA56705.2023.10100032
40. Kumar, R., Malhotra, R.K., Pandey, S., Gehlot, A., Gautam, I., Chamola, S.: Role of artificial intelligence in input tax credit reconciliation. In: Proceedings—2023 3rd International Conference on Pervasive Computing and Social Networking, ICPCSN 2023, pp. 497–501 (2023). https://doi.org/10.1109/ICPCSN58827.2023.00086
41. Kumar, V., Gupta, S., Khanna, R.: Dengue fever-a worldwide study. J. Med. Pharm. Allied Sci. **10**, 102–108 (2021). https://doi.org/10.22270/jmpas.vic2i1.2014
42. Malhotra, R.K., Sharma, C.A.P., Upreti, H., Sharma, S., Prakash, N., Mittal, S.: Role of psychological contract in employee engagement in higher education: empirical study on Uttarakhand. In: E3S Web of Conferences, vol. 556. EDP Sciences (2024). https://doi.org/10.1051/e3sconf/202455601030
43. Maurya, S.K., Ghosal, A., Manna, A.: Experimental investigations during fabrication and electrical discharge machining of hybrid Al/(SiC+ZrO2+NiTi) MMC. Int. J. Mach. Mach. Mater. **24**(3–4), 215–230 (2022). https://doi.org/10.1504/ijmmm.2022.125195
44. Miri, S.M., Farzaneh-Gord, M., Hosseinpour, A., Bajaj, M., Zaitsev, I.: The optimum sizing of zero-emission water-cooled VCR cycle based on exergo-economic-environmental assessment criteria by triple-objective MPSO. Sci. Rep. **14**(1) (2024). https://doi.org/10.1038/s41598-024-78994-z
45. Mittal, A., Pandey, H., Arsh, M., Soni, M., Goyal, H.R., Sharma, S.: Advancements and applications of artificial intelligence in various sectors: from internet of vehicles to 6G and healthcare. In: Proceedings—2024 4th International Conference on Pervasive Computing and Social Networking, ICPCSN 2024, pp. 107–112 (2024). https://doi.org/10.1109/ICPCSN62568.2024.00026
46. Nagpal, G., Singh, D.B.: Improving exergo-enviro-economic parameters and annual productivity of double slope solar desalting unit by incorporating concentrator integrated evacuated tubular collectors. Desalin. Water Treat. **306**, 9–21 (2023). https://doi.org/10.5004/dwt.2023.29796
47. Naik, B., Mishra, R., Kumar, V., Mishra, S., Gupta, U., Rustagi, S., Gupta, A.K., Preet, M.S., Bhatt, S.C., Rizwanuddin, S.: Micro-algae: revolutionizing food production for a healthy and sustainable future. J. Agric. Food Res. **15**,(2024). https://doi.org/10.1016/j.jafr.2023.100939
48. Özen, E., Singh, A., Taneja, S., Rajaram, R., Davim, J.P.: Innovating Sustainability Through Digital Circular Economy. IGI Global (2024). https://doi.org/10.4018/979-8-3373-0578-3
49. Özen, E., Singh, A., Taneja, S., Rajaram, R., Davim, J.P.: Sustainability, Innovation, and Consumer Preference. IGI Global (2024). https://doi.org/10.4018/979-8-3693-9699-5
50. Pandey, A., Chettri, N.: International conference on mountain ecosystems: biodiversity and adaptations under climate change scenario (22–24 March 2023). Biodiversity **24**(3), 112–113 (2023). https://doi.org/10.1080/14888386.2023.2210130
51. Pokhariya, H.S., Singh, D.P., Prakash, R.: Investigating the effect of urbanization on environmental quality using RS and GIS. In: 14th International Conference on Advances in Computing, Control, and Telecommunication Technologies, ACT 2023, vol. 2023-June, pp. 1459–1466. Grenze Scientific Society (2023)
52. Pramanik, A., Sinha, A., Chaubey, K.K., Hariharan, S., Dayal, D., Bachheti, R.K., Bachheti, A., Chandel, A.K.: Second-generation bio-fuels: strategies for employing degraded land for climate change mitigation meeting united nation-sustainable development goals. Sustainability (Switzerland), **15**(9) (2023). https://doi.org/10.3390/su15097578
53. Prasad, M.S.G., More, P., Christa, S., Kumar, M.A.: A chronic disease diagnosis model for smart healthcare systems enabled by artificial intelligence and the internet of things. In: Applications of Artificial Intelligence in the Healthcare Sector. Nova Science Publishers, Inc. (2023)

54. Rai, A., Sharma, V.K., Sharma, M., Singh, S.M., Singh, B.N., Pandey, A., Nguyen, Q.D., Gupta, V.K.: A global perspective on a new paradigm shift in bio-based meat alternatives for healthy diet. Food Res. Int. **169**,(2023). https://doi.org/10.1016/j.foodres.2023.112935
55. Raja, S., Agrawal, A.P., Patil, P., Thimothy, P., Capangpangan, R.Y., Singhal, P., Wotango, M.T.: Optimization of 3D printing process parameters of polylactic acid filament based on the mechanical test. Int. J. Chem. Eng. (2022). https://doi.org/10.1155/2022/5830869
56. Rajora, R., Gupta, H., Malhotra, S., Devliyal, S., Sunil, G.: Advancing renal health: unleashing the power of CNNs in multi-class classification for precise kidney condition diagnosis. In: 2024 IEEE 9th International Conference for Convergence in Technology, I2CT 2024 (2024). https://doi.org/10.1109/I2CT61223.2024.10544018
57. Raman, R., Buddhi, D., Lakhera, G., Gupta, Z., Joshi, A., Saini, D.: An investigation on the role of artificial intelligence in scalable visual data analytics. In: 2023 International Conference on Artificial Intelligence and Smart Communication, AISC 2023, pp. 666–670 (2023). https://doi.org/10.1109/AISC56616.2023.10085495
58. Ramanan, M., Singh, L., Kumar, A.S., Suresh, A., Sampathkumar, A., Jain, V., Bacanin, N.: Secure blockchain enabled cyber-physical health systems using ensemble convolution neural network classification. Comput. Electr. Eng. **101**,(2022). https://doi.org/10.1016/j.compeleceng.2022.108058
59. Ramesh, J.V.N., Khandekar, S.D., Baburao, D., Singh, M., Biban, L., Muniyandy, E., Desingurajan, N.: Harnessing remote sensing and geographic information systems to address soil degradation and water scarcity in climate change era. Remote Sens. Earth Syst. Sci. **7**(2), 139–147 (2024). https://doi.org/10.1007/s41976-024-00108-0
60. Rana, M.S., Dixit, A.K., Rajan, M.S., Malhotra, S., Radhika, S., Pant, B.: An empirical investigation in applying reliable industry 4.0 based Machine Learning (ML) approaches in analysing and monitoring smart meters using multivariate analysis of variance (Manova). In: 2022 2nd International Conference on Advance Computing and Innovative Technologies in Engineering, ICACITE 2022, pp. 603–607 (2022). https://doi.org/10.1109/ICACITE53722.2022.9823597
61. Rawat, S., Banerjee, D., Aggarwal, P., Singh, M.: Intelligent material extrusion defect prediction: CNN-SVM integration. In: 2024 International Conference on Advances in Modern Age Technologies for Health and Engineering Science, AMATHE 2024 (2024) https://doi.org/10.1109/AMATHE61652.2024.10582167
62. Samantaray, A., Chattaraj, S., Mitra, D., Ganguly, A., Kumar, R., Gaur, A., Mohapatra, P.K.D., Santos-Villalobos, S. de los, Rani, A., Thatoi, H.: Advances in microbial based bio-inoculum for amelioration of soil health and sustainable crop production. Curr. Res. Microb. Sci. **7** (2024). https://doi.org/10.1016/j.crmicr.2024.100251
63. Sehajpal, S., Kaur, N.: Neuroscience, nursing , and financial innovation : a synergistic approach to optimized patient care and economic well-being. In: Sanjay, T., Bhupinderpal Pal, S.C., Amar, J., Ercan, O., Pawan, K. (eds.) Neuromarketing's role in sustainable finance, pp. 133–146. IGI Global Scientific Publishing (2025). https://doi.org/10.4018/979-8-3693-9117-4.ch008
64. Semwal, P., Painuli, S., Shabaaz Begum, J.P., Jamloki, A., Rauf, A., Olatunde, A., Mominur Rahman, M., Mukerjee, N., Khalil, A.A., Aljohani, A.S.M., Al Abdulmonem, W., Simal-Gandara, J.: Corrigendum to "exploring the nutritional and health benefits of pulses from the indian himalayan region: a glimpse into the region's rich agricultural heritage" [Food Chem. 422 (2023) 136259] (Food Chemistry (2023) 422, (S0308814623008774). Food Chem. **445** (2024). https://doi.org/10.1016/j.foodchem.2024.138830
65. Shafiq, M., Bhavani, N.P.G., Venkata Naga Ramesh, J., Veeresha, R.K., Talasila, V., Sulaiman Alfurhood, B.: Thermal modeling and Machine learning for optimizing heat transfer in smart city infrastructure balancing energy efficiency and climate impact. Therm. Sci. Eng. Prog. **54** (2024). https://doi.org/10.1016/j.tsep.2024.102868
66. Sharma, B., Kaushal, D., Sharma, M., Joshi, S., Gopal, S., Gupta, P.: Integration of AI, digital twin and Internet of Medical Things (IoMT) for healthcare 5.0: a bibliometric analysis. In: 2023 International Conference on Advances in Computation, Communication and Information Technology, ICAICCIT 2023, pp. 1313–1318 (2023). https://doi.org/10.1109/ICAICCIT60255.2023.10466141

67. Sharma, M., Shah, J.K., Joshi, S.: Modeling enablers of supply chain decarbonisation to achieve zero carbon emissions: an environment, social and governance (ESG) perspective. Environ. Sci. Pollut. Res. **30**(31), 76718–76734 (2023). https://doi.org/10.1007/s11356-023-27480-6
68. Sharma, P., Malhotra, R.K., Altybekova, D., Singh, C.P.: AI and machine learning as catalysts for educational equity: bridging the urban–rural divide. Lect. Notes Netw. Syst. 1074 LNNS, 41–55 (2024). https://doi.org/10.1007/978-981-97-6103-6_4
69. Sharma, P., Malhotra, R.K., Kumar, R.: How is digitalization reshaping the landscape of Rural banking in India? In: 2024 International Conference on Smart Devices, ICSD 2024 (2024). https://doi.org/10.1109/ICSD60021.2024.10750981
70. Sharma, S., Gupta, C., Khanna Malhotra, R., Upreti, H.: Sustainable banking practices: impact, challenges and opportunities. In: E3S Web of Conferences, vol. 556. EDP Sciences (2024). https://doi.org/10.1051/e3sconf/202455601031
71. Sharma, S., Kandpal, V., Choudhury, T., Santibanez Gonzalez, E.D.R., Agarwal, N.: Assessment of the implications of energy-efficient technologies on the environmental sustainability of rail operation. AIMS Environ. Sci. **10**(5), 709–731 (2023). https://doi.org/10.3934/environsci.2023039
72. Sharma, S., Mishra, R.R., Joshi, V., Kour, K.: Analysis and interpretation of global air quality. In: 2020 11th International Conference on Computing, Communication and Networking Technologies, ICCCNT 2020 (2020). https://doi.org/10.1109/ICCCNT49239.2020.9225532
73. Shehzad, N., Ramtiyal, B., Jabeen, F., Mangla, S.K., Vijayvargy, L.: Metaverse adoption as a cornerstone for sustainable healthcare firms in the industry 5.0 epoch. J. Enterp. Inf. Manage. **37**(4), 1254–1281 (2024). https://doi.org/10.1108/JEIM-10-2023-0559
74. Shreya, S., Arya, P., Gupta, A.: Natural plant-derived bioactive compounds as health promoters. In: Microbial Bioactive Compounds: Industrial and Agricultural Applications (2023). Springer Nature. https://doi.org/10.1007/978-3-031-40082-7_9
75. Singh, A., Goel, A.: Supply chain optimization decisions for environmentally conscious products with preservation investment and carbon tax policy. Process Integr. Optim. Sustain. (2024). https://doi.org/10.1007/s41660-024-00467-7
76. Singh, A., Kumar, P., Taneja, S.: Artificial intelligence and machine learning-powered smart finance. In: Sanjay Taneja, P.K., Singh, A. (eds.) IGI Global. IGI Global (2024). https://doi.org/10.4018/979-8-3693-3264-1
77. Singh, A.P., De, K., Uniyal, V.P., Sathyakumar, S.: Unveiling of climate change-driven decline of suitable habitat for Himalayan bumblebees. Sci. Rep. **14**(1) (2024). https://doi.org/10.1038/s41598-024-52340-9
78. Singh, G., Aggarwa, R., Kumar, S., Nagpa, S., Dhondiyal, S.A.: Role of AI in enhancing health insurance penetration in rural India. In: Proceedings—2024 International Conference on Computational Intelligence and Computing Applications, ICCICA 2024, pp. 198–202. Institute of Electrical and Electronics Engineers Inc. (2024). https://doi.org/10.1109/ICCICA60014.2024.10584969
79. Singh, P., Gargi, B., Semwal, P., Verma, S.: Global research and research progress on climate change and their impact on plant phenology: 30 years of investigations through bibliometric analysis. Theoret. Appl. Climatol. **155**(6), 4909–4923 (2024). https://doi.org/10.1007/s00704-024-04919-5
80. Singh, S., Singh, A., Mohan, A., Batola, M., Kumar, N.: Health expenditures and health outcomes in central Europe and the Baltic region. Baltic Reg. **16**(2), 126–143 (2024). https://doi.org/10.5922/2079-8555-2024-2-7
81. Solanki, A., Singh, P., Pandit, M., Sawle, Y., Alotaibi, M.A., Malik, H., Márquez, F.P.G., Afthanorhan, A.: Techno-economic-environmental assessment of stand-alone hybrid renewable energy system for different batteries using HOMER-Pro. Int. J. Math. Eng. Manage. Sci. **9**(4), 779–800 (2024). https://doi.org/10.33889/IJMEMS.2024.9.4.040
82. Srivastava, S.P., Malhotra, R.K., Mukherjee, R.K.: Data driven decision making in insurance. In: 2024 International Conference on Smart Devices, ICSD 2024 (2024). https://doi.org/10.1109/ICSD60021.2024.10751068

83. Supreeth, B.R., Ansari, M.S. Al, Jasim, N.H., Swain, P.C., Srinivasan, D., Pundir, S.: Machine learning based corporate climate change disclosure in integrating institutional and resource approach based on computational intelligence method. In: Proceedings of 5th International Conference on Contemporary Computing and Informatics, IC3I 2022, pp. 2034–2038 (2022). https://doi.org/10.1109/IC3I56241.2022.10073204
84. Taneja, S., Chahal, B.P.S., Johri, A., Ozen, E., Kumar, P.: Neuromarketing's Role in Sustainable Finance. IGI Global (2024). https://doi.org/10.4018/979-8-3693-9117-4
85. Taneja, S., Gupta, S., Singh, A., Bhatnagar, M., Gaur, A.: Green management—a new paradigm in the world of business. Green Manage. New Paradig. World Bus. 1–230 (2023). https://doi.org/10.52305/DLIH6622
86. Taneja, S., Kumar, P., Reepu, Kukreti, M., Ozen, E.: Data Alchemy in the Insurance Industry (2024). Emerald Publishing Limited. https://doi.org/10.1108/978-1-83608-582-920241033
87. Taneja, S., Ozen, E., Kumar, P., Reepu.: E-Finance: Transformative Strategies for Sustainable Development in the Digital Era. Nova Science Publisher Inc. (2025). https://doi.org/10.52305/NZSJ1375
88. Thapa, D., Kumar, V., Naik, B., Kumar, V., Gupta, A.K., Mohanta, Y.K., Mishra, B., Rustagi, S.: Harnessing probiotic foods: managing cancer through gut health. Food Sci. Biotechnol. **33**(9), 2141–2160 (2024). https://doi.org/10.1007/s10068-024-01638-5
89. Thirumalaivasan, N., Nangan, S., Kanagaraj, K., Rajendran, S.: Assessment of sustainability and environmental impacts of renewable energies: focusing on biogas and biohydrogen (Biofuels) production. Process Saf. Environ. Prot. **189**, 467–485 (2024). https://doi.org/10.1016/j.psep.2024.06.063
90. Uniyal, A., Prajapati, Y.K., Ranakoti, L., Bhandari, P., Singh, T., Gangil, B., Sharma, S., Upadhyay, V.V., Eldin, S.M.: Recent advancements in evacuated tube solar water heaters: a critical review of the integration of phase change materials and nanofluids with ETCs. Energies **15**(23) (2022). https://doi.org/10.3390/en15238999
91. Upreti, H., Malhotra, R.K.: Bridging the urban-rural education gap in India through CSR (Corporate Social Responsibility) initiatives: a conceptual study with special reference to sustainable development goal 4 (quality education). In: E3S Web of Conferences, vol. 556. EDP Sciences (2024). https://doi.org/10.1051/e3sconf/202455601032
92. Upreti, H., Malhotra, R. K., Kumar, R., Karki, B.: Leveraging smart devices for corporate social responsibility in India: advancements, trends, and concerns towards sustainability. In: 2024 International Conference on Smart Devices, ICSD 2024 (2024). https://doi.org/10.1109/ICSD60021.2024.10751578
93. Verma, A., Sharma, D., Pant, M., Khatib, M.N., Singh, M.P., Zahiruddin, Q.S., Rustagi, S., Gaidhane, S., Gaidhane, A.M.: First sighting of human H5N1 in Australia: a detailed account and public health implications. New Microbes New Infect. 60–61 (2024). https://doi.org/10.1016/j.nmni.2024.101447
94. Verma, R., Kumar, G., Upadhyay, V.V., Ratur, A., Rao, A.L.N., Kumar, A., Boob, N.S.: Ecofriendly approaches in nanomaterial synthesis for sustainable healthcare applications. In: E3S Web of Conferences, vol. 511. EDP Sciences (2024). https://doi.org/10.1051/e3sconf/202451101024
95. Wazid, M., Singh, J., Das, A.K., Rodrigues, J.J.P.C.: An ensemble-based machine learning-envisioned intrusion detection in industry 5.0-driven healthcare applications. IEEE Trans. Consum. Electron. **70**(1), 1903–1912 (2024). https://doi.org/10.1109/TCE.2023.3318850
96. Yeruva, A.R., Vijaya Durga, C.S.L., Gokulavasan, B., Pant, K., Chaturvedi, P., Srivastava, A.P.: A smart healthcare monitoring system based on fog computing architecture. In: Proceedings of International Conference on Technological Advancements in Computational Sciences, ICTACS 2022, pp. 904–909 (2022). https://doi.org/10.1109/ICTACS56270.2022.9987881
97. Yildirim, S., Yildirim, D.C., Demirtaş, I., Kandpal, V.: Green transition impacts on the economy, society, and environment. In: Green Transition Impacts on the Economy, Society, and Environment. IGI Global (2024). https://doi.org/10.4018/979-8-3693-3985-5
98. Yuman, Y., Goyal, S.B., Rajawat, A.S., Kumar, M., Shankar, A., Alhayan, F., Basheer, S.: A blockchain-based solution for enhancing the efficiency and security of healthcare knowledge

management systems in the era of industry 4.0. Wirel. Netw. (2024). https://doi.org/10.1007/s11276-024-03770-w

99. Zaheer, A., Zehra Hussain, D., Verma, A., Sah, S., Pant, M., Nazli Khatib, M., Pratap Singh, M., Syed Zahiruddin, Q., Rustagi, S.: Marburg virus: An emerging threat to global health. Clin. Infect. Pract. **23** (2024). https://doi.org/10.1016/j.clinpr.2024.100371

Correction to: Investor Perception in the Digital Age: The Role of Deepfake Technology in Corporate Transparency

Bhushan Shrikant Joshi, Nilesh Gokhale, and Luan Vardari

Correction to:
Chapter 6 in: S. Taneja et al. (eds.), *Navigating the Deepfake Conundrum: A Manager's Roadmap*, Information Systems Engineering and Management 44, https://doi.org/10.1007/978-3-031-90742-5_6

In the original version of the book, the following belated corrections have been incorporated: The author name "Nilesh Gokhle" has been changed to "Nilesh Gokhale" in the Frontmatter and in Chapter 6. The book has been updated with the changes.

The updated version of this chapter can be found at
https://doi.org/10.1007/978-3-031-90742-5_6

S. Taneja et al. (eds.), *Navigating the Deepfake Conundrum: A Manager's Roadmap*,
Information Systems Engineering and Management 44,
https://doi.org/10.1007/978-3-031-90742-5_17

The manufacturer's authorised representative in the EU is Springer Nature Customer Service Centre GmbH, Europaplatz 3, 69115 Heidelberg, Germany. If you have any concerns regarding our products, please contact ProductSafety@springernature.com

Printed and bound by CPI Group (UK) Ltd, Croydon, CR0 4YY
30/09/2026
02225592-0012